Computer Logic

Computer Logic

Alan Rose

Reader in Mathematics
University of Nottingham

WILEY–INTERSCIENCE
a division of John Wiley & Sons Ltd.
London · New York · Sydney · Toronto

Library of Congress catalog card number 72-161088
ISBN 0 471 73510 8

Printed in Great Britain by John Wright & Sons Ltd.,
at The Stonebridge Press, Bristol

Preface

This book is intended for mathematicians who intend to work with computers, but who have no previous knowledge of Mathematical Logic. The propositional calculus is developed side by side with its practical applications to the theory of logical decision elements. Each new step in the propositional calculus is followed, as soon as possible, by a practical application. Many of these applications have, in the past, been discussed with reference to an algebraic model of the propositional calculus known as 'Boolean algebra' and the relationship between this algebra, the propositional calculus and decision elements is explained. Where it has seemed preferable, applications of Boolean algebra, rather than of the propositional calculus, have been made.

The different methods of construction of decision elements are discussed, including the comparatively modern method of using what are known as 'integrated circuits'. Decision elements can be combined, both to form the main part of a logical computer and to form the various components of a general-purpose digital computer, and these methods of use are discussed. Our discussions at this stage include the design of single-digit memory elements known as 'flip-flops'.

The concept of a 'designation number' of a Boolean expression is introduced, together with a related concept for the propositional calculus, and these concepts are applied to give methods of solution to problems of practical significance in computing, including the solution of simultaneous Boolean equations. Subsequently these equations are considered in connexion with further work on flip-flops and other problems involving the concept of memory, much use being made of the methods of Huffman and Mealey.

The University　　　　　　　　　　　　　　　　　　　　　　　A. Rose
Nottingham
England

Glossary of Symbols

$x, y, z, \ldots$ elementary mathematical variables
$p, q, r, \ldots$ propositional variables (alternatively $X, Y, Z, \ldots$)
$P, Q, R, \ldots$ syntactical variables (alternatively $\mathfrak{A}, \mathfrak{B}, \mathfrak{C}, \ldots$)

$-, \neg, \sim, N$	negation	
$\vee, A$	disjunction	
$\cdot, \&, K, \wedge$	conjunction	Alternative notations are shown
$\rightarrow, \supset, C$	implication (material)	here. The tilde symbol may be
$\dashv\!\!\rightarrow, \not\supset, B$	non-implication	used for the unary negation
$\leftrightarrow, \sim, \equiv, E$	equivalence (material)	functor or the binary equiva-
$\not\equiv, E'$	non-equivalence	lence functor (but not both).
$/, S$	incompatability	Square brackets are not used
$\downarrow, J$	joint denial	for any other purpose
$[\,,\,], D$	conditioned disjunction	

$=$ equals
$+$ addition (elementary or Boolean)
$\cup$ Boolean addition
$<$ is less than
T true
F false
$!$ factorial
$\Phi(P_1, \ldots, P_n)$ formula (other similar notations are used)
$=_\mathrm{T}$ equal in truth value
$\leqslant$ is less than or equal to
iff   if and only if
$\dashv 3$ strict implication
$>$ is greater than
$\geqslant$ is greater than or equal to
$F(\,, \ldots,\,)$ unspecified functor
Σ summation by addition or disjunction
$*$ asterisk (used to distinguish repeated formulae, also to denote translations of formulae and related Boolean variables)
Ł   Polish crossed L
ń   Polish accented n
$D_2(\,,\,), D_3(\,,\,,\,), \ldots$ disjunction and higher disjunctions
Π summation by multiplication or conjunction
$C_2(\,,\,), C_3(\,,\,,\,), \ldots$ conjunction and higher conjunctions
$=_\mathrm{df}$ equals by definition

$\equiv$ (mod 2) congruent, modulo 2 (other similar notations are used)

$\{A, B, C, D, E\}$ set whose members are A, B, C, D, E (other similar notations are used)

$=$ number of elements of the set

$?$ question mark (denotes freedom of choice of digit in designation number)

$\times$ cross (used in denoting the size of a matrix)

$\mathbf{x}, \mathbf{y}, ..., \mathbf{T}$ matrix variables

$|-|$ absolute difference

$\{,,\},$ special ternary functors (braces and angular brackets are

$\langle,,\rangle$ mostly used in the conventional manner)

$F_n(,...,)$ special functor (other similar notations are used)

t, f logical constants

$P^D, Q^D, ...$ duals

$.$ multiplication (elementary or Boolean)

$\cap$ Boolean multiplication

$'$ Boolean unary operation, interpretable as complement

$/$ division

$\in$ is a member of

$A, B, ...$ Boolean variables

Contents

Chapter 1

Logical Decision Elements

1.1 Propositional variables and connectives

The basic building bricks of computers consist largely of pieces of hardware (i.e. electrical mechanisms) known as 'decision elements'. Decision elements when used alone cannot form sequential mechanisms, but, as we shall see in later chapters, they form the main components of such mechanisms. These are best studied in connexion with the basic discipline of the science of Mathematical Logic; a discipline often known as the 'propositional calculus', alternative names being 'sentential calculus' and 'statement calculus'.

In elementary algebra it is customary to use an alphabet

$$x, y, z, \ldots$$

of letters to denote variable rational or real numbers and, in more advanced work, these are used to denote variable complex numbers, vectors, matrices, etc. We now extend this use to denote a variable proposition or sentence. Such sentences are considered as unanalysed wholes, no attempt at division, into subject and predicate or otherwise, being made.

We shall use a potentially infinite* alphabet of variables

$$p, q, r, \ldots$$

which we shall call propositional (or sentential) and which will denote variable propositions so that their possible values will be definite (true or false) propositions such as the following:

(a) Nottingham is in Ireland.
(b) $2+3 = 5$.
(c) $6 < 4$.
(d) Easter occurs in November.
(e) The sun sometimes shines in France.

Clearly the second and fifth of the above are true and the others are false.

---

* Although there are strictly only 26 letters in the alphabet $p, q, r, \ldots$ we may attach any number of primes to each letter to form a new symbol of the 'alphabet'.

Normally, in the propositional calculus, we may assign to a statement one of two truth values true (*T*), false (*F*).

In elementary algebra we may combine variables by operations such as addition, multiplication and (in the case where the variable denotes an integer) the operation of forming the factorial. Similarly, in the propositional calculus, we may apply logical operations to propositional variables to form new variable propositions and, as in elementary algebra, reapply the operations to the newly formed expressions as often as we wish. Most algebraic operations are binary, this being the case for the above examples of addition and multiplication, though we do sometimes use unary operations, as in the case of the factorial. We may also, on occasions, use operations which combine more than two expressions at a time as when, for instance, we construct

$$\max(x, y, z)$$

from x, y, z. Similarly most, but not all, of the logical operations which we shall consider are binary.

Two of the commonest logical operations which we shall cover are those of 'negation' and 'conjunction'. We shall denote these by the symbols $\sim$, & respectively. From propositional variables p, q we may construct the new (variable) propositions

$$\sim p \quad \sim q \quad p \,\&\, q$$

Thus negation is a unary operation while conjunction is binary. It should be noted that while p, q denote *arbitrary* statements the expressions $\sim p$, $\sim q, p \,\&\, q$ do not. The first two expressions denote variable propositions which must contain the word NOT and the third denotes a variable proposition which must contain the word AND. Having formed the above expressions we may continue to construct new ones such as

$$\sim p \,\&\, q \quad \sim(p \,\&\, q) \quad [\![\sim(\sim p \,\&\, q) \,\&\, \sim(p \,\&\, q)]\!] \,\&\, r$$

Any expression constructed in this way, provided that it is meaningful when its propositional variables are interpreted as meaningful statements, is known as a *formula* of the propositional calculus. The concept of a formula will be defined more precisely in the next chapter.

1.2 Decision elements considered as 'black boxes'. The AND truth table and the AND and NOT decision elements. Syntactical variables and truth tables for general formulae. Tautologies, absurdities and mixed formulae

In elementary algebra the value of the expression obtained when other expressions are combined by an operation is often required. Thus, for example, it may be necessary to know the value of the product xy when the values of x and y are known and, for this reason, multiplication tables are learned. Similarly, we may wish to know the truth value of a formula of the propositional calculus when we know the truth values of the formulae from which it is immediately constructed, provided that this construction amounts to combination by a given operation. If this is possible, we may set out the rules for the determination in the form of a table. Thus, for example, the truth value of the formula $p \& q$ may be read from the following table, provided that the truth values of p and q are known.

$p \& q$	T	F	q
T	T	F	
F	F	F	
p			

If p, q take as their values propositions which, in both cases, have the truth value T, then the corresponding statement $p \& q$ will also take the truth value T but, in the three remaining cases, it will take the truth value F.

We note that the truth value of $p \& q$ depends only on those of p and q (or, more precisely, on the truth values of the propositions which are the values of p and q) and not on their inner meanings. We say, therefore, that conjunction is a *truth-functional mode of composition* and refer to the above table as the truth table for conjunction. Similarly negation is truth-functional and has the truth table given below.

p	$\sim p$
T	F
F	T

We may, by repeated use of the truth tables of the logical operations (often known as 'functors') occurring in a formula, determine the truth value of that formula (as long as the truth values of all propositional variables occurring are known). Strictly speaking, however, our tables do not entitle us to say that, since the formula

$$p \& q$$

takes the truth value T when p, q both take the truth value T, then the formula

$$\sim (p \& q)$$

takes the truth value F when p, q both take the truth value T. This argument would depend on the fact that $\sim (p \& q)$ takes the truth value F when $p \& q$

takes the truth value T. The truth table for negation tells us that if the particular formula p takes the truth value T then its negation takes the truth value F; but it does not explicitly tell us anything about the negation of any other formula, even if it is the negation of another propositional variable. We could, of course, argue that as negation is a truth-functional mode of composition then, when the formulae

$$p \quad p\,\&\,q$$

both take the same truth value (T in this case), their negations

$$\sim p \quad \sim(p\,\&\,q)$$

must take the same, but this is an *inference*, not something stated in the table.

In order to ensure that the information required is stated more explicitly we introduce a new, potentially infinite, alphabet

$$P, Q, R, \dots$$

of variables known as *syntactical variables*. These represent arbitrary formulae of the propositional calculus, rather than statements which are in turn represented by the formulae. Thus the syntactical variables are not part *of* our propositional calculus, but part of the *syntax language* in which we talk *about* our propositional calculus. They enable us to discuss formulae without specifying exactly those to which we refer and we are thereby enabled to keep our discussion sufficiently general. The truth tables for conjunction and negation may be rewritten in terms of syntactical variables, so that the first table tells us the truth value of the conjunction of *any* two formulae (given the truth values of those formulae) since any two formulae whatever may be taken as the values of the syntactical variables P, Q. Similarly the second table tells us the truth value of the negation of any formula, given the truth value of that formula.

$P\,\&\,Q$	Q		$\sim P$
	T	F	
T	T	F	F
F	F	F	T
P			

Let us suppose, for example, that we wish to determine the truth value of

$$\sim[\![(p\,\&\,q)\,\&\sim r]\!]$$

given that the propositional variables p, q, r take the truth values T, T, F respectively. Since p, q both take the truth value T the conjunction table tells us, when we take p, q as values of P, Q respectively, that the sub-formula $p \& q$ takes the truth value T. Since r takes the truth value F the negation table tells us, when we take r as an instance of P, that $\sim r$ takes the truth value T. Taking $p \& q$, $\sim r$ as instances of P, Q respectively we see, by inspection of the conjunction table, that the sub-formula

$$(p \& q) \& \sim r$$

takes the truth value T. Finally, using the negation truth table and taking $(p \& q) \& \sim r$ as an instance of P, we see that the given formula takes the truth value F.

If p, q, r take the truth values T, F, T respectively we may infer, by a somewhat similar though shorter argument, that the formula $\sim [\![(p \& q) \& \sim r]\!]$ takes the truth value T. Since r takes the truth value T the sub-formula $\sim r$ takes the truth value F. If we now use the conjunction truth table and take $p \& q$, $\sim r$ as instances of P, Q respectively we see that, as $\sim r$ takes the truth value F and both the right-hand column entries in this table are F, the formula $(p \& q) \& \sim r$ takes the truth value F irrespective of the truth value of $p \& q$. There is therefore no need for us to evaluate the truth value of the sub-formula $p \& q$. We may say immediately that the sub-formula $(p \& q) \& \sim r$ takes the truth value F and, taking this sub-formula as an instance of P in the negation table, infer that the given formula takes the truth value T.

It is often convenient, when determining the truth value of a formula, to write the truth value of each propositional variable immediately below its every (relevant) occurrence, and the truth value of each of the remaining sub-formulae (where relevant) immediately under its principal connective (i.e. the connective used last in the construction of that sub-formula from propositional variables). Ultimately the truth value of the given formula (an improper sub-formula) is written under its principal connective. For example, the two earlier evaluations of the truth value of the formula $\sim [\![(p \& q) \& \sim r]\!]$ may be set out as follows:

$$\sim [\![(p \& q) \& \sim r]\!] \qquad \sim [\![(p \& q) \& \sim r]\!]$$
$$F \;\; TTT \;\; T\, TF \qquad T \qquad\quad F\, FT$$

In the second example we do not write the truth values under p, q or the first occurrence of the conjunction functor (the principal connective of the sub-formula $p \& q$) since the truth values of these three formulae are irrelevant.

Similarly if p, q take the truth values T, F respectively, we may establish, as shown below, that the formula

$$[\![\sim (p \& q) \& p]\!] \& \{ \sim q \& \sim [\![q \& (p \& q)]\!] \}$$

takes the truth value T. We shall use the convention that the negation functor binds more closely than the conjunction functor, so that the sub-formula

$$\sim q \,\&\, \sim [\![q \,\&\, (p \,\&\, q)]\!]$$

is an abbreviation for

$$\sim (q) \,\&\, \sim [\![q \,\&\, (p \,\&\, q)]\!]$$

and not for

$$\sim \{q \,\&\, \sim [\![q \,\&\, (p \,\&\, q)]\!]\}$$

The determination of the truth value of the formula may therefore be set out as follows:

$$[\![\sim (p \,\&\, q) \,\&\, p]\!] \,\&\, \{\sim q \,\&\, \sim [\![q \,\&\, (p \,\&\, q)]\!]\}$$
$$T \quad FF \; TT \quad T \quad TFTT \; FF$$

The truth values of the first and last occurrences of p and the last occurrence of q and $p \,\&\, q$ are not shown as they are irrelevant.

In general the truth value of a formula will depend on the truth values of the propositional variables occurring in it. This, of course, was found to be the case for $\sim [\![(p \,\&\, q) \,\&\, \sim r]\!]$ and such a formula is called a *mixed formula*. However, certain formulae are capable of taking only one truth value. A formula which always (in view of its logical structure) takes the truth value F is known as an *absurdity* and a formula which always takes the truth value T is known as a *tautology*. For example, $p \,\&\, \sim p$ is an absurdity and $\sim (p \,\&\, \sim p)$ is a tautology.

Let us now consider a piece of hardware with two input wires and an output wire, each being capable of two physical states, the same two states in all three cases. We shall not, in the present discussion, take into account the exact nature of these states or the method of construction of the hardware. We shall therefore regard the hardware simply as a 'black box' and refer to the physical states as 'state 1' and 'state 2'. The black box may, for example, be made from electromagnetic relays and metal rectifiers and state 1 (2) may

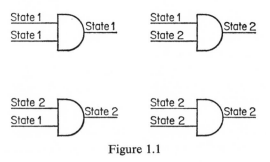

Figure 1.1

be connexion to the positive (negative) terminal of a direct current electricity supply, negative terminals being connected to earth. However, this and several other possible interpretations[1] of the present discussion will not be considered until later in the chapter.

Let us suppose that the output wire is in state 1 when both the input wires are also in this state and that, in the three remaining cases, the output wire is in state 2. Thus the state of the output wire depends on the states of the two input wires, but not on anything else, as shown by the following 'state table'.

	1	2	Bottom input
1	1	2	
2	2	2	

Top input

If we now identify the top and bottom inputs with the respective formulae P, Q and the states $1, 2$ with the truth values T, F respectively, we obtain the table given below.

	T	F	Q
T	T	F	
F	F	F	

P

This is, of course, the truth table of the formula $P \& Q$. We say, therefore, that this piece of hardware is *a decision element for* the conjunction functor and we may represent it diagrammatically as shown in Figure 1.2.

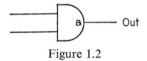

Figure 1.2

We make the convention that if n is a positive integer, the formulae $P_1, ..., P_n$ are combined by a functor of n arguments to form a new formula and the sub-formulae $P_1, ..., P_n$ occur in the new formula in such a way that, reading from left to right, P_j follows P_i whenever $1 \leqslant i < j \leqslant n$, then, for the decision element of this functor when used to simulate the new formula, the input corresponding to P_j occurs below the input corresponding to P_i whenever $1 \leqslant i < j \leqslant n$.

Thus the output of the device shown in Figure 1.3 would be considered to correspond to the formula $P \& Q$ rather than to $Q \& P$. In the present example

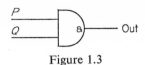

Figure 1.3

this point is not of very great importance since the conjunction functor satisfies the commutative law

$$P \,\&\, Q =_{\mathrm{T}} Q \,\&\, P$$

(cf. Examples 1A, 3(iii)). Let us, in order to obtain a less trivial example, consider the formula $P \,\&\, \sim Q$.

If a 'black box' has one input and one output, the state of the output wire always being opposite to that of the input wire, then the black box is a decision element for the negation functor. Since the two output and input states are compatible, we may connect the output of a negation decision element to an input of a conjunction element. If we choose the bottom input and if the top corresponds to a formula P, with the input to the negation decision element corresponding to a formula Q, then the top and bottom inputs to the conjunction decision element will correspond to the

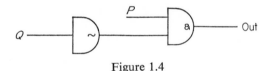

Figure 1.4

formulae P, $\sim Q$ respectively. The final output will then correspond to the formula $P \,\&\, \sim Q$ (cf. the last part of Section 5).

Examples 1A

1. Which of the following formulae are (a) tautologies, (b) absurdities, (c) mixed formulae?
   (i) $p \,\&\, \sim q$ (ii) $\sim [\![(p \,\&\, q) \,\&\, (r \,\&\, \sim p)]\!] \,\&\, \sim (q \,\&\, \sim q)$
   (iii) $(p \,\&\, q) \,\&\, \sim (p \,\&\, r)$ (iv) $(p \,\&\, q) \,\&\, (\sim p \,\&\, r)$
2. Prove that the formula

$$[\![(p \,\&\, q) \,\&\, \sim (p \,\&\, r)]\!] \,\&\, \sim (\sim r \,\&\, s)$$

   takes the truth value T if and only if p, q, r, s take the truth values T, T, F, F respectively.
3. Using the notation

$$P =_{\mathrm{T}} Q$$

   to denote that, under all assignments of truth values to the propositional variables occurring in the formulae denoted by the syntactical variables P, Q, the truth values of P, Q are the same, prove that

(i) $\sim \sim P =_T P$ (ii) $P \& P =_T P$ (iii) $P \& Q =_T Q \& P$
(iv) $(P \& Q) \& R =_T P \& (Q \& R)$ (v) $P \& \sim (Q \& \sim Q) =_T P$

4. Prove that if the formulae P, $\sim (P \& \sim Q)$ are tautologies then so is Q.
5. Prove that if P is a tautology then so is

$$\sim \{\langle \sim [\![(P \& Q) \& \sim R]\!] \& \sim (R \& \sim S) \rangle \& [\![Q \& \sim (P \& S)]\!] \}$$

6. Prove that if

$$\sim (\sim P \& \sim Q) =_T Q$$

and no propositional variable occurs both in P and in Q then P is an absurdity or Q is a tautology.

Solutions 1A

1. (i) $p \& \sim q$ $p \& \sim q$ mixed
   $TT\,TF$ FF

   (ii) $\sim [\![(p \& q) \& (r \& \sim p)]\!]$ tautology
   $TFF\,FT$
   $TF\,FF$

   $\sim (q \& \sim q)$ tautology
   $TF\,FT$
   $T\,FF$

   $\sim [\![(p \& q) \& (r \& \sim p)]\!] \& \sim (q \& \sim q)$ tautology
   TTT

   (iii) $(p \& q) \& \sim (p \& r)$ mixed
   $TTT\ TTFF$
   FFF

   (iv) $(p \& q) \& (\sim p \& r)$ absurdity
   $F\ FT\ F$
   FFF

2. If the given formula takes the truth value T then, since there is only one entry of T in the conjunction truth table,

$$(p \& q) \& \sim (p \& r) \quad \sim (\sim r \& s)$$

take the truth value T. Hence

$$p \& q \quad \sim (p \& r) \quad \sim r \& s$$

take the truth values T, T, F respectively and it then follows that

$$p \quad q \quad p \& r$$

take the truth values T, T, F respectively. Inspection of the conjunction truth table shows that, since $p, p \& r$ take the truth values T, F respectively, r takes the truth value F. Hence $\sim r$, $\sim r \& s$ take the truth values T, F respectively and s takes the truth value F.

Conversely

$$[\![(p \& q) \& \sim (p \& r)]\!] \& \sim (\sim r \& s)$$
$$TTTTTFF\ TTFF$$

3. (i) $\sim \sim P$ (ii) $P \& P$
$\quad\quad T\ F\ T \quad\quad\quad T\ T\ T$
$\quad\quad F\ T\ F \quad\quad\quad F\ F\ F$

(iii) $P \& Q$ takes the truth value T iff*
$\quad\quad P, Q$ take the truth value T iff
$\quad\quad Q \& P$ takes the truth value T.

(iv) $(P \& Q) \& R$ takes the truth value T iff
$\quad\quad P \& Q, R$ take the truth value T iff
$\quad\quad P, Q, R$ take the truth value T iff
$\quad\quad P, Q \& R$ take the truth value T iff
$\quad\quad P \& (Q \& R)$ takes the truth value T.

(v) $\sim (Q \& \sim Q)$ tautology $P \& \sim (Q \& \sim Q)$
$\quad\quad T \quad F F\ T \quad\quad\quad\quad\quad\quad T T T$
$\quad\quad T \quad F\ F \quad\quad\quad\quad\quad\quad\quad F F$

1.3 The truth table and decision element for equivalence. A relay circuit for this element

Another commonly used binary functor is the equivalence functor, for which we shall use the symbol $\equiv$. The formula $P \equiv Q$ may be read 'P is equivalent to Q' or, more precisely, 'P iff Q' or 'as a matter of fact, P is equivalent to Q'. The present equivalence functor does not correspond to the concept of inter-deducibility, as a strict equivalence functor with this property would not be truth-functional. Our (material) equivalence functor has the truth table given below, so that the formula $P \equiv Q$ takes the truth value T if and only if the truth values of P, Q are the same.

$P \equiv Q$	T	F	Q
T	T	F	
F	F	T	
P			

One method of making decision elements for this and other functors is by means of electromagnetic relays and, in some cases (though not in the present case of the equivalence functor), metal rectifiers, the physical states 1 and 2 now being interpreted as connexion to the positive and negative terminals respectively of a d.c. electricity supply. As engineering considerations make it convenient to earth negative terminals, connexions to the negative terminal will, in all cases considered here and in Section 1.7, be shown as earth connexions.

* 'iff' is used as an abbreviation for 'if and only if'.

The relays which we shall consider consist of a coil of wire enclosing a bar of iron, which is magnetized when a current is passed through the coil, but which loses its magnetism as soon as the current ceases to flow. The bar, when magnetized, attracts a hinged piece of metal against the action of a spring, thereby causing a two-way switch to change position, the new positions being, in Figure 1.5, denoted by broken lines. When the iron loses its magnetism the action of the spring causes the switch to return to its original position.

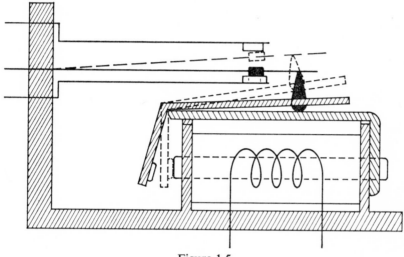

Figure 1.5

When drawing circuits for decision elements we shall normally depict only the coil and the switch. The circuit shown in Figure 1.6 (which is essentially that of McCallum and Smith[2]) provides a decision element for the equivalence functor. The switch is, as in all our relay circuits, shown in the position which it occupies when no current flows through the coil and the ends of the coil are connected to inputs whose physical states (connexion to the positive or to earth) correspond to the truth values of P and Q.

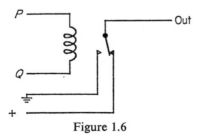

Figure 1.6

If P, Q take the same truth value then both ends of the coil are connected to the same terminal of the electricity supply, and no current passes through the coil. The iron therefore remains unmagnetized so that the switch does not change over, but retains the position shown in Figure 1.6. Thus the output wire is connected to the positive terminal of the electricity supply and the physical state of the output wire corresponds to the truth value T, i.e. to the truth value of the formula $P \equiv Q$. If, on the other hand, the formulae P, Q take opposite truth values, then the two ends of the coil are connected to different electrical terminals and a current flows through the coil, causing the iron to become magnetized. Thus the switch changes over and the output wire is connected to earth, the state of this wire therefore corresponding to the truth value F, i.e. the truth value now taken by the formula $P \equiv Q$.

1.4 The other binary truth tables

Let us suppose, for the time being, that n is a positive integer and that we wish, from general* formulae $P_1, ..., P_n$, to construct (using all of these n formulae but no others), a formula $\Phi(P_1, ..., P_n)$ whose truth value is, in every possible case, determined by the truth values of $P_1, ..., P_n$. We may therefore set up a truth table for this formula $\Phi(P_1, ..., P_n)$ and we say that this is *of degree n*. Thus, for example, the truth tables for $\sim P$, $P \& Q$, $P \equiv Q$ are of degrees $1, 2, 2$ respectively.

Since each of the formulae $P_1, ..., P_n$ may take one of two truth values, we may assign these values to $P_1, ..., P_n$ in exactly 2^n ways and, in each of the 2^n cases, the truth table will give the value of the formula $\Phi(P_1, ..., P_n)$. Since the answer to the question 'What is the truth value of $\Phi(P_1, ..., P_n)$?' must be given in exactly 2^n cases and, in each of these 2^n cases, the answer must be given in one of exactly two possible ways, there are exactly 2^{2^n} ways of giving all the answers. Thus there are exactly 2^{2^n} truth tables of degree n.¹

In particular, letting $n = 1$, we see that there are exactly 2^{2^1} or 4 truth tables of degree 1 and, similarly, that there are exactly 16 truth tables of degree 2.

Two of the 4 first-degree truth tables are those of P (trivially) and $\sim P$ (ignoring the suffix of P_1) and the other two are of formulae which are a tautology and an absurdity. Thus the negation table is the only truth table of degree 1 which we shall find to be of much interest.

In the case $n = 2$ we shall, similarly, find that only 8 (or perhaps 10) of the 16 truth tables are of much interest. Apart from the 2 binary tables (for

---

* We make no assumptions about the nature of the formulae $P_1, ..., P_n$. Thus we must consider every possible assignment of truth value to them, the choice of truth value for a particular P_i ($1 \leqslant i \leqslant n$) not being restricted by the choices of truth values of the others.

conjunction and equivalence) which we have considered already, our main interest will be in the tables for disjunction, implication, non-implication, non-equivalence, incompatibility and joint denial. The following notations will be used:

$$
\text{For} \left\{
\begin{array}{l}
P \text{ or } Q \\
P \text{ implies (materially) } Q \text{ (or if } P \text{ then } Q) \\
P \text{ does not imply } Q \text{ (or } P \text{ but not } Q) \\
P \text{ is not equivalent to } Q \\
P \text{ is incompatible with } Q \text{ (or not both} \\
\quad P \text{ and } Q) \\
\text{neither } P \text{ nor } Q
\end{array}
\right\}
\text{ we write}
\left\{
\begin{array}{l}
P \vee Q \\
P \supset Q \\
P \not\supset Q \\
P \not\equiv Q \\
P/Q \\
\\
P \downarrow Q
\end{array}
\right\}
$$

$$
\text{We call the functor}
\left\{
\begin{array}{l}
\vee \\
\supset \\
\not\supset \\
\not\equiv \\
/ \\
\downarrow
\end{array}
\right\}
\text{ the functor of}
\left\{
\begin{array}{l}
\text{disjunction (or alternation)} \\
\text{material implication} \\
\text{non-implication} \\
\text{non-equivalence} \\
\text{incompatibility} \\
\text{joint denial}
\end{array}
\right\}
$$

Let us now consider the truth tables of these functors. Unless otherwise stated, 'or' will refer to the inclusive form of disjunction, corresponding to the Latin 'vel' rather than 'aut'. Thus $P \vee Q$ takes the truth value T if and only if at least one of P, Q takes the truth value T and the disjunction table is as follows:

$P \vee Q$	T	F	Q
T	T	T	
F	T	F	
P			

The notation '$P \supset Q$' denotes the statement 'if P then Q' or 'as a matter of fact, P implies Q' rather than 'P strictly implies Q' or 'Q may be deduced from P'. The strict implication functor, which is normally denoted by the symbol $\dashv3$, is not a truth-functional mode of composition. This may be seen by considering the statements

(i) Grass is red strictly implies that all trains have square wheels.

(ii) $2 > 3$ strictly implies $3 > 4$.

In both cases we may consider the formula

$$p \dashv3 q$$

noting that, in both cases, the value of p is a false statement, as is the value

of q. However, in the first case, the value of $p \rightarrow 3q$ is a false statement since
the propositions 'Grass is red' and 'All trains have square wheels' are not
connected in any way while, in the second case, the value of $p \rightarrow 3q$ is a true
statement, since we may infer '3 > 4' from '2 > 3' by adding 1 to each side of
the inequality. The statement 'if P then Q' may reasonably be regarded as
true, except when P, Q take the truth values T, F respectively and we therefore
adopt the table given below.

$P \supset Q$	T	F	Q
T	T	F	
F	T	T	
P			

For instance, a university might require (among other things) that applicants
for admission must have a G.C.E. pass in English or Latin and that, if they
wish to study a scientific subject after admission, they must have a G.C.E.
pass in Mathematics. Thus, if the values of p, q, r, s are 'the candidate has
passed in English', 'the candidate has passed in Latin', 'the candidate has
passed in Mathematics', 'the candidate wishes to study a scientific subject
after admission' respectively, the condition that a candidate satisfies these
requirements is

$$(p \lor q) \,\&\, (s \supset r)$$

In the case of a candidate wishing to study History and with English as his
only G.C.E. pass, the truth values of p, q, r, s are T, F, F, F respectively and
the condition is, as we show below, satisfied. If the lower right-hand entry in
the $\supset$

$$(p \lor q) \,\&\, (s \supset r)$$
$$T T F \; T \quad F T F$$

truth table had been F the condition would not, of course, have been satisfied.

Perhaps the most important property of the material implication functor is
the property that, if the formulae P, Q are constructed in such a way that, for
all assignments of truth values to the propositional variables occurring in P
and Q under which P takes the truth value T, Q takes the truth value T
(Q possibly taking the truth value T under some additional assignments), then
the formula $P \supset Q$ is a tautology.

$$P \supset Q$$
$$T T T$$
$$F T T$$
$$F T F$$
$$T \;\; F \quad \text{(by hypothesis, this case does not arise)}$$

The truth table of the non-implication functor is, as might be expected from its name, determined by the equation

$$P \not\supset Q =_{\text{T}} \sim(P \supset Q)$$

$P \supset Q$	T	F	Q
T	F	T	
F	F	F	
P			

Clearly, by inspection of this table,

$$P \not\supset Q =_{\text{T}} P \mathbin{\&} \sim Q$$

so the non-implication functor may be read BUT NOT.

The truth table of the non-equivalence functor is, of course, determined by the equation

$$P \not\equiv Q =_{\text{T}} \sim(P \equiv Q)$$

$P \not\equiv Q$	T	F	Q
T	F	T	
F	T	F	
P			

Thus the formula $P \not\equiv Q$ takes the truth value T if and only if exactly one of the formulae P, Q takes the truth value T and non-equivalence may be regarded as the EXCLUSIVE OR. However, as stated above, OR normally refers to the inclusive form and the functor $\not\equiv$ is normally known as 'non-equivalence'.

The truth table of the INCOMPATIBILITY or NOT BOTH functor is determined by the equation

$$P/Q =_{\text{T}} \sim(P \mathbin{\&} Q)$$

while that of the JOIN T DENIAL functor is determined by the equation

$$P \downarrow Q =_{\text{T}} \sim P \mathbin{\&} \sim Q$$

P/Q	T	F	Q
T	F	T	
F	T	T	
P			

$P \downarrow Q$	T	F	Q
T	F	F	
F	F	T	
P			

We have now discussed, out of the 16 theoretically existing binary functors, the 8 non-trivial functors denoted by the symbols

$$\vee \quad \& \quad \supset \quad \not\supset \quad \equiv \quad \not\equiv \quad / \quad \downarrow$$

It should be noted, in order to simplify the calculations of truth values, that some rows and columns of the truth tables for 6 of these functors have two identical entries (cf. Section 1.2 and Solutions 1B, 2). Of the remaining 8, the only 2 of appreciable interest are converse implication and converse non-implication. As might be expected, the truth tables of these functors, which we shall denote by the symbols $\subset$, $\not\subset$ respectively, are determined by the equations

$$P \subset Q =_{\mathrm{T}} Q \supset P \qquad\qquad P \not\subset Q =_{\mathrm{T}} Q \not\supset P$$

$P \subset Q$	T	F	Q
T	T	T	
F	F	T	
P			

$P \not\subset Q$	T	F	Q
T	F	F	
F	T	F	
P			

The remaining 6 binary functors, which we shall denote by $F_i(,)$ $(i = 1, ..., 6)$ have truth tables such that $F_1(P, Q), F_2(P, Q)$ always take the truth values T, F respectively and

$$F_3(P, Q) =_{\mathrm{T}} P \quad F_4(P, Q) =_{\mathrm{T}} \sim P \quad F_5(P, Q) =_{\mathrm{T}} Q \quad F_6(P, Q) =_{\mathrm{T}} \sim Q$$

In Section 1.2 we made the convention that $\sim$ was to be regarded as a 'stronger' symbol than &, so that certain brackets could, by way of abbreviation, be omitted. We now extend this convention, so that & is regarded as a stronger symbol than $\vee$ and $\vee$ as a stronger symbol than any of the symbols for the remaining binary functors (of which 8 are trivial). Thus, for example, the formula

$$p \,\&\sim q \vee r \supset \sim r \vee s \,\&\,(q \equiv p)$$

is an abbreviation for

$$[\![(p \,\&\sim q) \vee r]\!] \supset \{\sim(r) \vee [\![s \,\&\,(q \equiv p)]\!]\}$$

the remaining five pairs of brackets being understood.

Examples 1B

1. (a) Abbreviate, by removing unnecessary brackets, the following formulae:
    (i) $[\![(p \vee q) \,\&\,(r \,\&\, s)]\!] \supset \{\langle[\![\sim(p) \vee q]\!] \,\&\, r\rangle \vee q\}$
    (ii) $[\![p/(q \equiv r)]\!] \vee (p \,\&\, q)$
    (b) Insert all 'understood' brackets in the following formulae:
    (i) $(p \vee \sim q \,\&\, r) \vee s \supset (p/\sim q \,\&\, r)$
    (ii) $\sim p \,\&\, q \vee r \not\equiv q$

2. Which of the following formulae are (a) tautologies, (b) absurdities, (c) mixed formulae?

   (i) $(p \supset q) \supset [\![\sim q \supset (p \supset r)]\!]$

   (ii) $(p \supset q) \supset \{(q \equiv r) \supset [\![(q \supset p) \supset (p \equiv r)]\!]\}$

   (iii) $(p \& q) \& r \equiv [\![r \supset (p/q)]\!]$

   (iv) $[\![(p \downarrow q) \downarrow r]\!] \supset [\![p \downarrow (q \downarrow r)]\!]$

   (v) $[\![(p \not\equiv q) \not\equiv r]\!] \equiv p \& (\sim q \& \sim r) \vee [\![(\sim p \& q) \& \sim r \vee (\sim p \& \sim q) \& r]\!]$

   (vi) $(p \supset q) \supset [\![(r \supset s) \supset (p \& r \supset q \& s)]\!]$

   (vii) $(p \equiv q \vee r) \equiv (p \equiv q) \vee (p \equiv r)$

3. Prove that

   (i) $P \vee Q =_{\mathrm{T}} Q \vee P$

   (ii) $(P \vee Q) \vee R =_{\mathrm{T}} P \vee (Q \vee R)$

   (iii) $P \& (Q \vee R) =_{\mathrm{T}} P \& Q \vee P \& R$

4. Prove that, if Q takes the truth value F,

$$\sim P =_{\mathrm{T}} P \equiv Q$$

Use this equation to design a decision element for the negation functor.

5. Design a decision element for the non-equivalence functor.

6. (a) Abbreviate, where possible,

   (i) $(p \vee q) \supset \{[\![(p \vee r) \& \sim (q)]\!] \vee \sim (p \supset s)\}$

   (ii) $[\![(p \equiv q) \equiv (p \vee r)]\!] \supset [\![(p \not\equiv r) \vee q]\!]$

  (b) Rewrite, without using abbreviations,

   (i) $p \vee (q \& \sim r \vee s) \supset q \& p \vee \sim r$

   (ii) $[\![(p \supset q) \supset \sim r \vee q]\!] \equiv (p \vee q \equiv r)$

7. Which of the following formulae are (a) tautologies, (b) absurdities, (c) mixed formulae?

   (i) $p \vee q \& r \supset (p \vee q) \& r$

   (ii) $\{[\![(q \supset r) \supset (p \supset r)]\!] \supset s\} \supset [\![(p \supset q) \supset s]\!]$

   (iii) $(p \equiv q \vee r) \supset [\![(p \equiv q) \vee (p \equiv r)]\!]$

   (iv) $[\![(p \equiv q) \vee (p \equiv r)]\!] \supset (p \equiv q \vee r)$

   (v) $\{[\![(p \vee q) \& (p \supset q)]\!] \& (q \supset p)\} \& (p/q)$

   (vi) $(p \equiv q \& r) \supset (p \equiv q) \& (p \equiv r)$

8. Prove that

   (i) $P \equiv P \& Q =_{\mathrm{T}} P \supset Q$

   (ii) $Q \equiv (P \not\supset Q) =_{\mathrm{T}} P \downarrow Q$

   (iii) $P \supset Q \vee R =_{\mathrm{T}} (P \supset Q) \vee (P \supset R)$

   (iv) $P \supset (Q \supset R) =_{\mathrm{T}} Q \supset (P \supset R)$

   (v) $P \vee Q \& R =_{\mathrm{T}} (P \vee Q) \& (P \vee R)$

9. Prove that, using only non-equivalence and the logical constant t (i.e. a formula of constant truth value T), a formula having the same truth table as $\sim P$ can be constructed. Use this construction to design an alternative form of negation decision element.

Solutions 1B

1. (a) (i) $(p \vee q) \& (r \& s) \supset (\sim p \vee q) \& r \vee q$

     (ii) $[\![p/(q \equiv r)]\!] \vee p \& q$

  (b) (i) $\{\langle p \vee [\![\sim (q) \& r]\!]\rangle \vee s\} \supset \{p/[\![\sim (q) \& r]\!]\}$

     (ii) $\{[\![\sim (p) \& q]\!] \vee r\} \not\equiv q$

2.   (i) $(p \supset q) \supset \llbracket \sim q \supset (p \supset r) \rrbracket$ tautology
          T T FT
          T F TT
        TF FT

  (ii) $(p \supset q) \supset \{(q \equiv r) \supset \llbracket (q \supset p) \supset (p \equiv r) \rrbracket\}$ tautology
     TF FT
       T T TF FT
       T TF FT
       T FF TT
       T T T TT T
       T T T FT F
     $(p \neq_T q$ or $q \neq_T r$ or $p =_T q =_T r)$

  (iii) $(p \& q) \& r \equiv \llbracket r \supset (p/q) \rrbracket$ absurdity
         F FF FT

  If r takes the truth value T
     $(p \& q) \& r \equiv \llbracket r \supset (p/q) \rrbracket =_T p \& q \equiv (p/q) =_T (p \& q) \equiv \sim (p \& q)$
                                 T F F T
                                   F F T F

  (iv) $\llbracket (p \downarrow q) \downarrow r \rrbracket \supset \llbracket p \downarrow (q \downarrow r) \rrbracket$ mixed
           FT T
        TF TF F TF

  (v) $\llbracket (p \neq q) \neq r \rrbracket \equiv p \& (\sim q \& \sim r) \lor (\llbracket \sim p \& q) \& \sim r \lor (\sim p \& \sim q) \& r \rrbracket$ mixed
     T T FT T TT T T FTT F T
     T FTT T F F FTF F FT F F F FT F F

  (vi) $(p \supset q) \supset \llbracket (r \supset s) \supset (p \& r \supset q \& s) \rrbracket$ tautology
          T T F F T
          T T FFT
       TF FT
          T TF FT
          T T T TT T

     (If p, r both take the truth value T we consider, in succession, the cases where (a) q takes the truth value F, (b) s takes the truth value F and (c) q, s both take the truth value T, ignoring irrelevant information.)

  (vii) $(p \equiv q \lor r) \equiv (p \equiv q) \lor (p \equiv r)$ mixed
       TT TT T TT T T
       FF TT F FT F T

3.   (i) $P \lor Q, Q \lor P$ take the truth value F iff P, Q also take the truth value F.

  (ii) $(P \lor Q) \lor R$ takes the truth value F iff
     $P \lor Q, R$ take the truth value F iff
     P, Q, R take the truth value F iff
     $P, Q \lor R$ take the truth value F iff
     $P \lor (Q \lor R)$ takes the truth value F

  (iii) If P takes the truth value T then
     $P \& Q =_T Q$ $P \& R =_T R$

  Thus
     $P \& (Q \lor R) =_T Q \lor R =_T P \& Q \lor P \& R$
     $P \& (Q \lor R) =_T P \& Q \lor P \& R$
     F F F F F F F

4. If Q takes the truth value F then the truth value of $P \equiv Q$ is determined by the right-hand column of the truth table and $P \equiv Q =_T \sim P$. Thus if, in the circuit for the $\equiv$ decision element, we replace the input corresponding to Q by an input corresponding to a false proposition, we obtain a decision element for negation. (The circuit is essentially that of McCallum and Smith.[2])

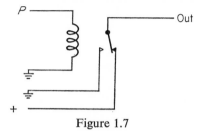

Figure 1.7

5. Since $P \not\equiv Q =_T \sim (P \equiv Q)$, the output of the decision element for non-equivalence must always be in the physical state opposite to the corresponding state for equivalence. Thus[2] we have only to reverse the positive and earth connexions of the circuit for the decision element for equivalence (Figure 1.8).

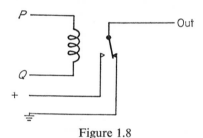

Figure 1.8

1.5 Combination of decision elements to form decision mechanisms for general formulae

In Section 1.2 we considered a particular case of a mechanism whose output corresponded to the truth value of a formula $(P \& \sim Q)$ containing more than one functor. More generally we may, given a formula $\Phi(P_1, ..., P_n)$ containing k occurrences of (not necessarily distinct) functors (other than occurrences within the sub-formulae $P_1, ..., P_n$), construct, from the corresponding k decision elements, a mechanism with n inputs whose states correspond to the truth values of the formulae $P_1, ..., P_n$ and whose output corresponds to the truth value of the formula $\Phi(P_1, ..., P_n)$. Such a mechanism will be referred to as a decision mechanism for the formula $\Phi(P_1, ..., P_n)$.

If the principal connective of the latter formula is a functor of i arguments we may use a decision element for this functor, connecting the i inputs to the outputs of the decision mechanisms for the i principal sub-formulae (i.e. the formulae combined by the principal connective) of $\Phi(P_1, ..., P_n)$. After a finite number of repetitions of the process we shall require only the inputs corresponding to $P_1, ..., P_n$.

More formally we may prove our result by a 'strong induction' method. Let the number, l, of (not necessarily distinct) symbols (other than brackets) occurring in $\Phi(P_1, ..., P_n)$ be known as the *length* of the formula $\Phi(P_1, ..., P_n)$. Thus, for example, the formula

$$(p \lor q) \lor \sim p$$

is of length 6 since the numbers of occurrences of $p, q, \lor, \sim$ are $2, 1, 2, 1$ respectively. We shall prove the result for the special case $l = 1$ (the lowest meaningful value of l) and then show that, if the result is true for all positive integers less than l, then it is also true for l. This type of induction is sometimes known as 'course of values induction' and is, of course, somewhat different from the conventional type of induction used early in Section 2.4. We shall assume that $\Phi(P_1, ..., P_n)$ is constructed entirely from functors and (some* or all of) the sub-formulae $P_1, ..., P_n$. We shall also, for the present, assume that $P_1, ..., P_n$ denote propositional† variables.

If $l = 1$ then, for some integer $w (1 \leqslant w \leqslant n)$, $\Phi(P_1, ..., P_n)$ is P_w and we need only a wire (Figure 1.9). (This step is known as the 'basis' of the induction.)

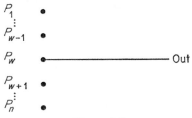

Figure 1.9

We now assume the result for $1, ..., l-1$ and deduce it for l. (This step is known as the 'induction step'.) Since we have disposed of the case $l = 1$ we may, in future, assume that $l \geqslant 2$ and therefore that the formula $\Phi(P_1, ..., P_n)$ has a principal connective $F(, ...,)$ of $i (i \geqslant 1)$ arguments. Thus $\Phi(P_1, ..., P_n)$ is of the form

$$F[\![\Psi_1(P_1, ..., P_n), ..., \Psi_i(P_1, ..., P_n)]\!]$$

---

* The insertion of the words 'some or' makes the statement of the result unnecessarily general, but it simplifies the induction step.

† We must still use syntactical variables, in order to ensure that our proof applies to *all* propositional variables. If we were to replace $P_1, ..., P_n$ by $p_1, ..., p_n$ respectively our proof would not apply to the propositional variables $q_1, ..., q_n$.

where no propositional variables, other than those denoted by $P_1, ..., P_n$, occur in the formula $\Psi'_v(P_1, ..., P_n)$ $(v = 1, ..., i)$ and

$$l(\Phi) = 1 + \sum_{v=1}^{i} l(\Psi'_v)$$

Thus

$$l(\Psi'_v) < l(\Phi) \quad (v = 1, ..., i)$$

and the induction hypothesis applies to the formula $\Psi'_v(P_1, ..., P_n)$ $(v = 1, ..., i)$. Hence, by this hypothesis, we may construct decision mechanisms for the formulae $\Psi'_v(P_1, ..., P_n)$ $(v = 1, ..., i)$ and, connecting the i outputs of these mechanisms to the inputs of a decision element for the functor $F(, ...,)$, we obtain the required decision mechanism (see Figure 1.10). Since all functors

Figure 1.10

are truth-functional the above construction remains valid even if $P_1, ..., P_n$ do not all denote propositional variables.

As an example of this procedure let us consider the construction of a decision mechanism for the formula

$$(p \vee q) \& \sim (p \supset r) \equiv (p \supset r)/\sim \sim q$$

The principal sub-formulae of the given formula are

$$(p \vee q) \& \sim (p \supset r) \quad (p \supset r)/\sim \sim q$$

Repeating the process for these formulae we obtain, as principal sub-formulae

$$p \vee q \quad \sim (p \supset r) \quad p \supset r \quad \sim \sim q$$

Further repetitions of the process give the following formulae, all those obtained at any one stage being printed on the same line and formulae which are repetitions of formulae already obtained being starred:

$$p \quad q \quad p \supset r^* \quad p \quad r \quad \sim q$$
$$p^* \quad r^* \quad q^*$$

Working from formulae on a line to those on the next, starting with the given formula and continuing until there are no unstarred formulae, we obtain the circuit shown in Figure 1.11. (A black dot at the junction of two wires indicates electrical contact.) The second occurrence of the sub-formula $p \supset r$ does not necessitate the use of a further decision element for material implication, as a second wire can be connected to the output wire of the first decision element. Since the implication functor is not commutative, the equation

$$P \supset Q =_T Q \supset P$$

being invalid, it is particularly important that the input

$$P \supset Q \qquad Q \supset P$$
$$T F F \qquad F T T$$

to the implication decision element corresponding to p should be drawn above that corresponding to r (cf. the last part of Section 1.2). Diagrams such as Figure 1.11 for circuits (not showing all the circuitry details) are known as 'block diagrams'.

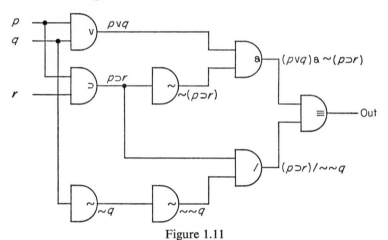

Figure 1.11

1.6 The notation of Łukasiewicz and other alternative notations

Perhaps the notation for the propositional calculus which differs most fundamentally from that used in this book is the notation of Łukasiewicz.[3] Before discussing this we shall mention briefly some other alternative notations.[4]

Usual notation in this book	Alternative notations
$p, q, r, \ldots, P, Q, \ldots$	$X, Y, Z, \ldots, \mathfrak{A}, \mathfrak{B}$
$P \& Q$	$P \cdot Q,\ \mathfrak{A} \& \mathfrak{B}$
$P \supset Q$	$P \rightarrow Q$
$P \not\supset Q$	$P -\!\!\mid\!\rightarrow Q$
$P \equiv Q$	$P \sim Q,\ P \leftarrow\!\mid\!\rightarrow Q$
$P \not\equiv Q$	$P \leftarrow\!\mid\!\rightarrow Q$
$\sim P$	$\bar{P},\ \neg\, P$
$p \& q$	$X \& Y$

Obvious alternatives (such as $\mathfrak{A} \cdot \mathfrak{B}$ for $P \& Q$ or $\mathfrak{A} \rightarrow \mathfrak{B}$ for $P \rightarrow Q$) are omitted.

The notation of Łukasiewicz uses, as we used above, alphabets

$$p, q, r, \ldots \quad P, Q, R, \ldots$$

of propositional variables and syntactical variables respectively, but it uses other capital Roman letters to denote functors and the letter for the principal connective of a sub-formula is always written as the first (reading from left to right) symbol of that sub-formula. The following table correlates the main notation of this book with the notation of Łukasiewicz.

For	$P \vee Q$	$P \& Q$	$P \supset Q$	$P \not\supset Q$	$P \equiv Q$	$P \not\equiv Q$	P/Q	$P \downarrow Q$	$\sim P$
Read	APQ	KPQ	CPQ	BPQ	EPQ	$E'PQ$	SPQ	JPQ	NP

The symbols

$$A, K, C, E, N$$

are used virtually always, but our other connectives are less standard. The notation has the advantage that it is easy to type or print and that all brackets are avoided.

If, for example, we wish to rewrite the formula

$$(p \vee q) \,\&\, r \supset \sim (p \equiv \sim q)/(q \not\supset p)$$

we must write first the symbol C (since the above formula has material implication as its principal connective) followed by the translation of the formula $(p \vee q) \,\&\, r$ and then by the translation of the formula

$$\sim (p \equiv \sim q)/(q \not\supset p)$$

If we denote the translation of a formula P by P^* we obtain, successively,

$$C[\![(p \vee q) \,\&\, r]\!]^* [\![\sim (p \equiv \sim q)/(q \not\supset p)]\!]^*$$

$$CK(p \vee q)^* rS[\![\sim (p \equiv \sim q)]\!]^* (q \not\supset p)^*$$

$$CKApqrSN(p \equiv \sim q)^* Bqp$$

$$CKApqrSNEp(\sim q)^* Bqp$$

$$CKApqrSNEpNqBqp$$

Similarly, if we now use stars to denote translations in the opposite

direction, we may, in successive stages, translate, as follows, the formula
AKpNEqNNCprSE′pqJNCprq.

$$(KpNEqNNCpr)^* \vee (SE′pqJNCprq)^*$$

$$p \, \& \, (NEqNNCpr)^* \vee [\![(E′pq)^*/(JNCprq)^*]\!]$$

$$p \, \& \sim (EqNNCpr)^* \vee \{(p \not\equiv q)/[\![(NCpr)^* \downarrow q]\!]\}$$

$$p \, \& \sim [\![q \equiv (NNCpr)^*]\!] \vee \{(p \not\equiv q)/[\![\sim (Cpr)^* \downarrow q]\!]\}$$

$$p \, \& \sim [\![q \equiv \, \sim (NCpr)^*]\!] \vee \{(p \not\equiv q)/[\![\sim (p \supset r) \downarrow q]\!]\}$$

$$p \, \& \sim [\![q \equiv \, \sim \sim (Cpr)^*]\!] \vee \{(p \not\equiv q)/[\![\sim (p \supset r) \downarrow q]\!]\}$$

$$p \, \& \sim [\![q \equiv \, \sim \sim (p \supset r)]\!] \vee \{(p \not\equiv q)/[\![\sim (p \supset r) \downarrow q]\!]\}$$

When carrying out the above translation we note first that the initial symbol *A* must be followed by exactly two meaningful formulae, so we must find a formula *Q* whose first symbol is the symbol *K* which follows this *A*. Thus *K* must be followed by two formulae of which the first is *p*, so the last symbol of *Q* will be the last symbol of the formula beginning with the first symbol (from left to right) *N*, and therefore of the formula beginning with the first *E*, the second or third *N* or the first *C*. This symbol *C* is the principal connective of the sub-formula *Cpr*, so the formula *Q* is *KpNEqNNCpr*. This formula *Q* is the first principal sub-formula of the given formula. Thus the other principal sub-formula of the given formula must be formed by the totality of the symbols occurring to the right of the sub-formula *Cpr* and this principal sub-formula is *SE′pqJNCprq*. Thus the first stage of the above work was to enter the expression

$$(KpNEqNNCpr)^* \vee (SE′pqJNCprq)^*$$

Similar arguments then led us ultimately to obtain, as the translation, the formula

$$p \, \& \sim [\![q \equiv \, \sim \sim (p \supset r)]\!] \vee \{(p \not\equiv q)/[\![\sim (p \supset r) \downarrow q]\!]\}$$

Examples 1C

1. Translate into the notation of Łukasiewicz
    (i) $(p \supset q \vee r \, \& \, s) \equiv \, \sim [\![\sim p \not\equiv (q \downarrow p)]\!]$
    (ii) $p \vee q \, \& \, r \not\supset [\![q \equiv \, \sim (p \vee r)]\!]$
2. Translate into the customary notation
    (i) *CKpKCpqCNrNqNEpNE′rq*
    (ii) *AKKApqNSrsqr*
3. Draw block diagrams for decision mechanisms corresponding to the following formulae:
    (i) $(p \supset q \, \& \, r) \equiv (q \vee s \not\supset q \, \& \, r)$
    (ii) *KE′pqCApNrJqNEpNs*
    (iii) *CSpCqNrEKAJpqrsNCqs*

4. Translate into the notation of Łukasiewicz
   (i) $(p \equiv q) \vee \llbracket r \,\&\, s \not\supset \sim(q \vee r) \rrbracket$
   (ii) $(p \supset q) \not\equiv \sim(p/q \vee r)$

5. Translate into the customary notation
   (i) $KKKEpNAqrKpqJpNAqpSEpqNNr$
   (ii) $EApKqNSrpCEpqBNEprNEqs$

Solutions 1C

1. (i) $ECpAqKrsNE'NpJqp$
   (ii) $BApKqrEqNApr$

2. (i) $p \,\&\, \llbracket (p \supset q) \,\&\, (\sim r \supset \sim q) \rrbracket \supset \sim \llbracket p \equiv \sim(r \not\equiv q) \rrbracket$
   (ii) $\llbracket (p \vee q) \,\&\, \sim(r/s) \rrbracket \,\&\, q \vee r$

3.

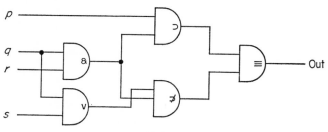

Solution 1C, 3(i)

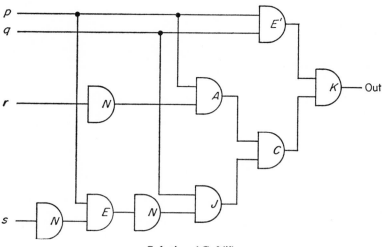

Solution 1C, 3(ii)

1.7 The practical construction of decision elements. The other circuits of McCallum and Smith and a simplification

We have considered, so far, the construction of relay circuits, using positive and earth connexions as the physical states corresponding to T and F respectively, of decision elements for negation and two non-trivial binary functors. We shall now consider the construction, by similar methods (involving also the use of rectifiers) of decision elements for the remaining six non-trivial binary functors. In the circuit shown in Figure 1.12 an arrow will be used to denote a rectifier which can pass current in the direction* in which the arrow points, but not in the opposite direction. Theoretically the rectifier has resistances of zero and infinity in the respective directions, but in practice departure from these ideal values causes limitations to the number of rectifiers which may be used in parallel in the circuits corresponding to conjunction and disjunction.

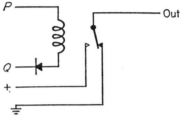

Figure 1.12

The output will be connected to the positive (and thus be in the physical state corresponding to the truth value T) if the states of the top and bottom inputs correspond to the truth values T, F respectively, since the rectifier does not prevent current from flowing from top to bottom. In the three remaining cases no current will flow through the coil, either because one terminal of the electricity supply is disconnected or because the rectifier prevents current flowing through the coil from bottom to top. Thus, in these three cases, the physical state of the output will correspond to F. Hence we see that the truth table

$P \not\Rightarrow Q$	T	F	Q
T	F	T	
F	F	F	
P			

---

* By convention, current is considered to flow from the positive terminal of the electricity supply to the negative, although the electrons do, in fact, move in the opposite direction.

of the functor corresponding to this decision element is the non-implication truth table. If we interchange the fixed connexions of the positive and negative terminals we obtain, in all four cases, an output opposite to that obtained above, so that we now have a decision element for implication, the circuit of Figure 1.13 (though not that of Figure 1.12) being essentially that

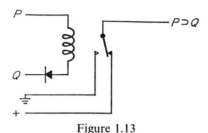

Figure 1.13

of McCallum and Smith. (Circuits in this section not acknowledged to McCallum and Smith are essentially as published by J. E. Parton and the author.[5])

Let us consider now the construction of a decision element for disjunction. In the circuit of Figure 1.14, it is impossible for current to flow from the

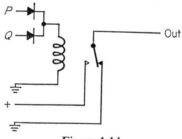

Figure 1.14

input terminal corresponding to the formula P to that corresponding to Q, since the bottom rectifier prevents this. Similarly the top rectifier prevents the flow of current in the reverse direction. Thus a necessary and sufficient condition for a current to flow through the coil (and therefore for the state of the output wire to correspond to T) is that the state of at least one input wire shall correspond to the truth value T. In the one case where the condition is not satisfied the position of the switch remains as shown in the diagram and the state of the output wire corresponds to the truth value F. Thus this circuit (which is based on a circuit of McCallum and Smith[2]), is that of a decision element for disjunction.

More generally, if, corresponding to the analogy of disjunction with addition (cf. Examples 1B, 3), we define the summation operator Σ by the following equations ($=_{\mathrm{df}}$ meaning 'equals by definition')

$$\sum_{i=\alpha}^{\alpha} P_i =_{\mathrm{df}} P_\alpha$$

$$\sum_{i=\alpha}^{\beta+1} P_i =_{\mathrm{df}} \left(\sum_{i=\alpha}^{\beta} P_i \right) \vee P_{\beta+1} \quad (\beta = \alpha, \alpha+1, \ldots)$$

and then define the truth tables of the functors $D_k(\,,\ldots,\,)$ of k arguments ($k = 2, 3, \ldots$) by the equations

$$D_k(P_1, \ldots, P_k) =_{\mathrm{T}} \sum_{i=1}^{k} P_i \quad (k = 2, 3, \ldots)$$

we may generalize the above circuit to provide circuits for decision elements for the functors $D_k(\,,\ldots,\,)$ ($k = 2, 3, \ldots$) as shown in Figure 1.15. Practical

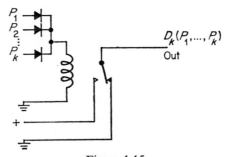

Figure 1.15

considerations, however (see above), would limit the value of k to about 8. For higher values of k we could, of course, make use of the fact that

$$D_{\alpha k}(P_1, \ldots, P_{\alpha k}) =_{\mathrm{T}} D_\alpha(D_k(P_1, \ldots, P_k), D_k(P_{k+1}, \ldots, P_{2k}), \ldots,$$
$$D_k(P_{(\alpha-1)k+1}, \ldots, P_{\alpha k})) \quad (k = 2, 3, \ldots; \alpha = 2, 3, \ldots)$$

Both the above formulae take the truth value F iff $P_1, \ldots, P_{\alpha k}$ take the truth value F.

For example, if $\alpha = 4$, $k = 6$ we may simulate a decision element for the functor $D_{24}(\,,\ldots,\,)$ as shown in Figure 1.16.

Methods similar to those used previously for non-equivalence and implication lead at once to the circuit shown in Figure 1.17 for a decision element for the joint denial functor, since

$$P \downarrow Q =_{\mathrm{T}} {\sim} (P \vee Q)$$

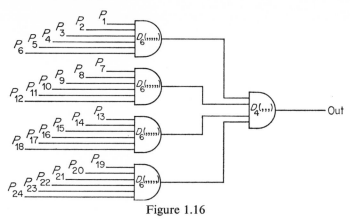

Figure 1.16

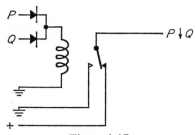

Figure 1.17

The circuit of McCallum and Smith for a conjunction decision element used resistances in addition to relays and rectifiers, but the circuit shown below is a more straightforward analogue of the circuits for the functors $D_k(, \dots,)$ $(k = 2, 3, \dots)$. Corresponding to the analogy between conjunction and multiplication (cf. Examples 1A, 3 (iii), (iv); 1B, 3 (iii)) we define the summation operator $\prod$ by

$$\prod_{i=\alpha}^{\alpha} P_i =_{\mathrm{df}} P_\alpha$$

$$\prod_{i=\alpha}^{\beta+1} P_i =_{\mathrm{df}} \left(\prod_{i=\alpha}^{\beta} P_i \right) \& P_{\beta+1} \quad (\beta = \alpha, \alpha+1, \dots)$$

Thus, if we define the truth tables of the functors $C_k(, \dots,)$ of k arguments $(k = 2, 3, \dots)$ by the equations

$$C_k(P_1, \dots, P_k) =_{\mathrm{T}} \prod_{i=1}^{k} P_i \quad (k = 2, 3, \dots)$$

a decision element for & may be constructed by the circuit shown in Figure 1.18 and, more generally,* the circuit of Figure 1.19 provides a decision element for the functor $C_k(\ ,...,\ )$ $(k = 2, 3, ...)$.

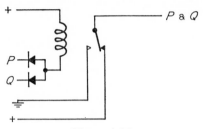

Figure 1.18

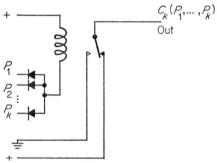

Figure 1.19

If any of the inputs corresponding to $P_1, ..., P_k$ is earthed (i.e. if any of $P_1, ..., P_k$ takes the truth value F) then a current flows through the coil and the output becomes earthed, so that its physical state corresponds to the truth value (F) of the formula $C_k(P_1, ..., P_k)$. In the remaining case neither end of the coil is earthed, so the output is (as shown in Figure 1.19) connected to the positive terminal. Thus the state of the output corresponds to the truth value (T) of $C_k(P_1, ..., P_k)$. As in the case of the functors $D_k(\ ,...,\ )$, the rectifiers inhibit the flow of current between two input terminals.

For large values of k there are, of course, practical difficulties, but we may, by methods similar to those discussed above for the functors $D_{\alpha k}(\ ,...,\ )$, make use of the equations

$$C_{\alpha k}(P_1, ..., P_{\alpha k}) =_T C_\alpha(C_k(P_1, ..., P_k), C_k(P_{k+1}, ..., P_{2k}), ...,$$
$$C_k(P_{(\alpha-1)k+1}, ..., P_{\alpha k}))\quad (k = 2, 3, ...; \alpha = 2, 3, ...)$$

---

* We note that $C_2(P_1, P_2) =_T P_1 \ \& \ P_2$.

A decision element for incompatibility may, by methods similar to those used above, be shown to be provided by the circuit of Figure 1.20, since

$$P/Q =_\text{T} {\sim}(P \mathbin{\&} Q)$$

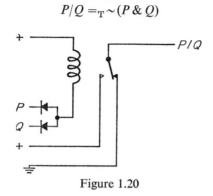

Figure 1.20

Examples 1 D

1. Draw a block diagram for a decision mechanism which simulates the functor $C_{42}(\,,\,...,\,)$, using only decision elements for the functors $C_i(\,,\,...,\,)$ ($i = 6, 7$).
2. Prove that, in the circuit shown below, the output corresponds to the formula $[P, Q, R]$, where $[P, Q, R] =_\text{T} P \mathbin{\&} Q \vee R \mathbin{\&} {\sim} Q$

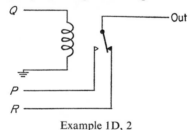

Example 1D, 2

3. Prove that

$$[t, P, Q] =_\text{T} P \vee Q$$

and deduce that a decision element for the functor $D_3(\,,\,)$ may be constructed, using a relay and only two rectifiers.
4. Design a decision element for the functor $C_3(\,,\,)$ using a relay and only two rectifiers.
5. Design a decision element for the implication functor, using a relay, but no rectifiers.
6. If $\langle P, Q, R \rangle =_\text{T} {\sim}[P, Q, {\sim}R]$ show how to design a decision element for the functor $\langle\,,\,,\,\rangle$ using two rectifiers and a double-wound, relay the switch being operated if at least one of the two relay coils is energized.
7. Show how to construct a decision element for conjunction with one relay, without using rectifiers.

8. Find a formula of the propositional calculus corresponding to the output of the decision element shown below if the switch cannot change over unless both relay coils are energized. (Currents flowing in the two coils cause opposite magnetization effects if and only if they flow in the same direction, i.e. both right to left or both left to right.)

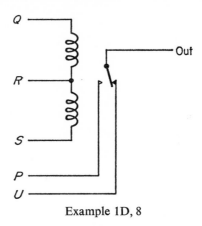

Example 1D, 8

Solutions 1 *D*

1.

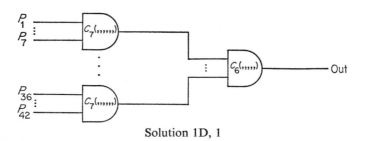

Solution 1D, 1

2. If Q takes the truth value T then current flows through the coil and the output wire is connected to the input corresponding to P. Otherwise the output is connected as shown in the diagram for Solution 1D, 1.

3. If P takes the truth value T then

$$[t, P, Q] =_T t =_T P \vee Q$$

If P takes the truth value F then

$$[t, P, Q] =_T Q =_T P \vee Q$$

Hence

$$D_3(P, Q, R) =_T [t, P \vee Q, R]$$

Thus, in the disjunction circuit, we replace the upper permanent connexion to earth by a connexion to the input corresponding to R. If $P \vee Q$ takes the truth value *F* then, as before, the switch is as shown in the diagram and the output is connected to the input corresponding to *R*. The output then corresponds to the truth value of *R* (and therefore to the truth value of $D_3(P, Q, R)$). If $P \vee Q$ takes the truth value *T* then, as before, the output corresponds to the truth value *T* (which is the truth value of $D_3(P, Q, R)$).

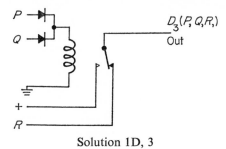

Solution 1D, 3

4. By an argument similar to that of the previous solution, we obtain the circuit shown below, using the equation

$$C_3(P, Q, R) =_T [R, P \& Q, f]$$

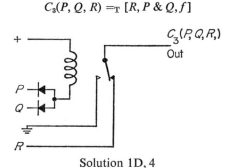

Solution 1D, 4

5.
$$P \supset Q =_T [Q, P, t]$$

Hence, by arguments similar to those used above, we obtain the circuit shown below.

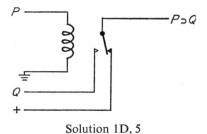

Solution 1D, 5

1.8 Duality and its applications. Conditioned disjunction and mixers

The theory of duality in the propositional calculus bears some analogy with the theory of duality in geometry. In plane projective geometry, for example, we may reverse the roles of points and lines, thereby obtaining new theorems. For instance the theorem 'If A and B are two distinct points then there exists a unique line l containing A and B' has, as its dual, the theorem 'If a and b are two distinct lines then there exists a unique point L contained in both lines'. Similarly we shall consider now the duality between the two truth values T, F.

If we consider the conjunction truth table and, for all eight entries of truth values in the table, replace each occurrence of T by F and each occurrence of F by T we obtain the second of the three truth tables shown below. Apart from the unusual order in which the entries are made in the new table, it is,

$P \& Q$	T	F	Q
T	T	F	
F	F	F	
P			

	F	T	Q
F	F	T	
T	T	T	
P			

$P \vee Q$	T	F	Q
T	T	T	
F	T	F	
P			

of course, the truth table of the disjunction functor. We may therefore say that the disjunction functor is the dual functor of conjunction or that the latter truth table is the dual of the former.

Similarly, if the syntactical variables $P_1, ..., P_n$ denote propositional variables and $\Phi(P_1, ..., P_n)$, $\Psi(P_1, ..., P_n)$ are two formulae, each of which contains these n propositional variables and no others, such that the truth table of $\Psi(P_1, ..., P_n)$ is obtained from that of $\Phi(P_1, ..., P_n)$ by interchanging, in all the $(n+1)2^n$ cases,* the truth values T, F then the formula $\Psi(P_1, ..., P_n)$ is said to be a dual of the formula $\Phi(P_1, ..., P_n)$.

For example, the formula $r \not\supset p \& q$ is a dual of the formula $p \vee q \supset r$.

p	q	r	$p \vee q \supset r$
T	T	T	T
F	T	T	T
T	F	T	T
F	F	T	T
T	T	F	F
F	T	F	F
T	F	F	F
F	F	F	T

p	q	r	$r \not\supset p \& q$
F	F	F	F
T	F	F	F
F	T	F	F
T	T	F	F
F	F	T	T
T	F	T	T
F	T	T	T
T	T	T	F

The above conditions are, of course, satisfied if and only if

$$\Psi(P_1, ..., P_n) =_\text{T} \sim \Phi(\sim P_1, ..., \sim P_n)$$

or

$$\Psi(\sim P_1, ..., \sim P_n) =_\text{T} \sim \Phi(P_1, ..., P_n)$$

Thus, if P, Q denote propositional variables, the formula $P \vee Q$ is a dual of $P \& Q$ and, by a similar argument, $P \& Q$ is a dual of $P \vee Q$. More generally, for any formulae P, Q, if Q is a dual of P then P is a dual of Q. The formula $P \vee Q$ is not a dual of $P \& Q$ for any choices whatever of P and Q, as shown by the following example, in which $p \& q$, $p \vee \sim r$ are taken as instances of P, Q respectively and p, q, r take the respective truth values (i) T, T, T, (ii) F, F, F.

$$(p \& q) \& (p \vee \sim r) \qquad p \& q \vee (p \vee \sim r)$$
$$T T \; T \; T \;\; T T \qquad\quad T \;\;\; T T F$$

However, if we replace every functor of a formula by the dual functor, we obtain† a dual formula. Since the negation functor is self-dual we may, in the case of the formula $(p \& q) \& (p \vee \sim r)$ considered above, obtain a dual

Dual truth table

P	$\sim P$		P	
T	F		F	T
F	T		T	F

* In general, truth tables must be set out as shown on this page so that each of the 2^n assignments of truth values corresponds to n entries corresponding to the truth values of $P_1, ..., P_n$ and one further entry corresponding to the truth value of $\Phi(P_1, ..., P_n)$.

† This will be proved shortly.

formula by interchanging all occurrences of conjunction and disjunction functors. Under assignment (ii) (above) the formula $p \& q \vee (p \vee \sim r)$ took the truth value T (from which we deduced that it was not a dual of the given formula), but the particular dual constructed by the present method takes the truth value F.

$$(p \vee q) \vee p \& \sim r$$
$$F F F \; F F F$$

Let us now prove the general rule stated above, for obtaining a dual of a formula. Let $P_1, ..., P_n$ denote propositional variables and let $\Phi(P_1, ..., P_n)$ be a formula containing all these n propositional variables and no others. Let $\Psi(P_1, ..., P_n)$ be the formula obtained from $\Phi(P_1, ..., P_n)$ by replacing each functor with a dual functor. We shall prove the result* by strong induction on the length, l, of $\Phi(P_1, ..., P_n)$.

If $l = 1$ then $n = 1$ and $\Phi(P_1, ..., P_n)$ is P_1, as is $\Psi(P_1, ..., P_n)$. Since $\sim \Phi(\sim P_1, ..., \sim P_n)$ is $\sim \sim P_1$ the result is

$$P_1 =_T \sim \sim P_1$$

which is trivial.

We now assume the result for $1, ..., l-1$ and deduce it for l. Let the principal connective of $\Phi(P_1, ..., P_n)$ be the functor $F(, ...,)$ of k arguments and let the dual functor by which it is replaced in the construction of $\Psi(P_1, ..., P_n)$ be $G(, ...,)$. Thus, by hypothesis, for all formulae $Q_1, ..., Q_k$,

$$G(Q_1, ..., Q_k) =_T \sim F(\sim Q_1, ..., \sim Q_k) \tag{A}$$

Since $\Phi(P_1, ..., P_n)$ is of the form

$$F[\![\Lambda_1(P_{11}, ..., P_{1j_1}), ..., \Lambda_k(P_{k1}, ..., P_{kj_k})]\!]$$

where

$$\{P_{i1}, ..., P_{ij_i}\} \subseteq \{P_1, ..., P_n\} \quad (i = 1, ..., k)$$

$$l(\Phi) = 1 + \sum_{i=1}^{k} l(\Lambda_i)$$

and it follows at once that

$$l(\Lambda_i) < l(\Phi) \quad (i = 1, ..., k)$$

Hence, by the induction hypothesis, setting

$$n = j_i, \quad P_w = P_{iw} \quad (w = 1, ..., j_i; \; i = 1, ..., k),$$

$$\Omega_i(P_{i1}, ..., P_{ij_i}) =_T \sim \Lambda_i(\sim P_{i1}, ..., \sim P_{ij_i}) \quad (i = 1, ..., k) \tag{B}$$

---

* More precisely we shall prove, by strong induction on l, that, for all n, $P_1, ..., P_n$ the result holds.

where $\Omega_i(P_{i1}, ..., P_{ij_i})$ is the formula obtained from $\Lambda_i(P_{i1}, ..., P_{ij_i})$ by replacing each functor with the same dual functor as used at the corresponding step in the construction of $\Psi(P_1, ..., P_n)$. Hence

$$\Psi(P_1, ..., P_n) = G(\Omega_1(P_{11}, ..., P_{1j_1}), ..., \Omega_k(P_{k1}, ..., P_{kj_k}))$$

$$=_T \sim F(\sim \Omega_1(P_{11}, ..., P_{1j_1}), ..., \sim \Omega_k(P_{k1}, ..., P_{kj_k})) \quad \text{(by A)}$$

$$=_T \sim F(\Lambda_1(\sim P_{11}, ..., \sim P_{1j_1}), ..., \Lambda_k(\sim P_{k1}, ..., \sim P_{kj_k})) \quad \text{(by B)}$$

$$= \sim \Phi(\sim P_1, ..., \sim P_n)$$

The remaining 8 non-trivial functors form pairs of duals as follows:

$$\supset, \not\subset \quad \subset, \not\supset \quad \equiv, \not\equiv \quad /, \downarrow$$

and this is easily checked from the truth tables. Another example of a self-dual functor[6] is provided by conditioned disjunction.[7] If P^D, Q^D, R^D are duals of P, Q, R respectively then a dual of $[P, Q, R]$ is $[R^D, Q^D, P^D]$. It will be sufficient, in order to prove this, to show that

$$[R, Q, P] =_T \sim [\sim P, \sim Q, \sim R]$$

If Q takes the truth value T then $\sim Q$ takes the truth value F and

$$[R, Q, P] =_T R =_T \sim \sim R =_T \sim [\sim P, \sim Q, \sim R]$$

If Q takes the truth value F then $\sim Q$ takes the truth value T and

$$[R, Q, P] =_T P =_T \sim \sim P =_T \sim [\sim P, \sim Q, \sim R]$$

A decision element for conditioned disjunction is sometimes known as a 'mixer' since the output will be in the same physical state as the first or third input according as the second input is in the first or second physical state. Of course, if the roles of the two states are reversed, the output will be in the same physical state as the first or third input according as the second input is in the second or first physical state. This result is obvious on physical grounds, but it is also a consequence of our duality result for conditioned disjunction since a decision element for a functor becomes a decision element for the dual functor when the roles of the two physical states are interchanged.

Examples 1E

1. Construct duals of the following formulae:
   (i) $p \vee q \ \& \ r \equiv [\![(p/q) \supset \sim (s \vee r)]\!]$
   (ii) $(p \downarrow q) \not\supset (p \equiv \sim r)$
2. Prove that if the formula $\{P, Q, R\}$ takes the truth value T if and only if at least two of the formulae P, Q, R take the truth value T then the functor $\{ , , \}$ is self-dual.
3. Design a decision element for the functor $\{ , , \}$.

4. If the truth table of the functor $F_n(\ ,\ ...,\ )$ of $2n-1$ arguments is such that the formula $F_n(P_1, ..., P_{2n-1})$ takes the truth value T if and only if at least n of the formulae $P_1, ..., P_{2n-1}$ take the truth value T $(n = 2, 3, ...)$, prove that the functor $F_n(\ ,\ ...,\ )$ is self-dual.

5. Prove that if, with the notation of question 2,

$$\langle P, Q, R \rangle =_T \{P, \sim Q, R\}$$

then the first functor is self-dual and that

$$P \vee Q =_T \langle P, f, Q \rangle$$

Use the duality result to establish the corresponding identity for conjunction.

Solutions 1E

1. (i) $p\ \&\ (q \vee r) \not\equiv [\![(p \downarrow q) \not\subset \sim (s\ \&\ r)]\!]$
    (ii) $(p/q) \subset (p \not\equiv \sim r)$

2. $\{P, Q, R\}$ takes the truth value T iff at least two of P, Q, R take the truth value T iff at most one of P, Q, R takes the truth value F iff at most one of $\sim P, \sim Q, \sim R$ takes the truth value T iff $\{\sim P, \sim Q, \sim R\}$ takes the truth value F iff $\sim\{\sim P, \sim Q, \sim R\}$ takes the truth value T. Thus

$$\{P, Q, R\} =_T \sim\{\sim P, \sim Q, \sim R\}$$

3. $\{P, Q, R\} =_T [P, P \equiv Q, R]$

Hence we may use the circuit shown below (which, as might be expected from the result of question 2, contains no fixed connexions to terminals of the supply of electricity).

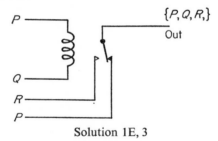

Solution 1E, 3

1.9 Practical constructions (continued). Transistors, integrated circuits and magnetic cores

Logical decision elements used in large computers have, in the past, been made from thermionic valves, transistors and magnetic cores. In recent years use of both thermionic valves and magnetic cores has decreased, largely because of the phenomenal technological advances which have occurred in transistor manufacture. These have given the latter such overwhelming advantages in terms of size, reliability and cost that they are now used exclusively. In particular the ability to construct several transistors, together with associated components such as diodes and resistors in a monolithic form no larger than a pin-head has resulted in 'integrated circuits'* providing the building blocks of all modern computers. It is not necessary for an understanding of computer logic to appreciate the physical processes occurring within a transistor. Indeed it is becoming increasingly unnecessary even to understand the use of transistors in logical decision elements as the variety and complexity of available integrated circuits expand. They may be quite adequately treated as 'black boxes'.

However, a brief description of transistor decision elements may be of interest and will be given here, but for a fuller understanding of the physical properties of transistors and their applications refer to Ledley.[9]

In order to understand the action of the transistor decision element, let us consider first a different application of the metal rectifiers referred to previously. Then they were used to control the direction of current flow through the coil of a magnetic relay. An alternative use is control over whether a current does or does not flow through a resistor. Note that the magnitude of the current is unimportant, as long as it is sufficient to allow the two states to be distinguished. This is most conveniently achieved by examining the voltage across the resistor.

Let us, in the first instance, represent the truth value x by a voltage level V_x ($x = T, F$), where $V_T > V_F$. For fairly small values of the positive integer k we may construct a decision element for the functor $D_k(\,,\dots,\,)$ (cf. Section 1.7) as shown in Figure 1.21.

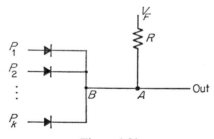

Figure 1.21

If the voltage at one or more of the input points P_i ($1 \leqslant i \leqslant k$) is V_T then current will flow from the input P_i through the (low resistance) rectifier and the resistance R to the point connected to the lower voltage level V_F. The rectifiers corresponding to input points at voltage level V_F may be ignored as, for currents flowing from the right to the left of the diagram, their

* For an account of the construction of integrated circuits see Khambata,[8] Chapter 5.

resistances are very high. As the rectifier resistance is low in comparison with the resistance R, the voltage level at the output point will be approximately V_T. If, on the other hand, the input points corresponding to $P_1, ..., P_k$ are all at voltage level V_F, no current will flow and there will be no voltage difference across the resistance R. Thus the output will also be at voltage V_F.

In computers, semiconductor rectifiers (often referred to as diodes) have long been used in place of the metal type because of their smaller physical size, but the operation is identical as far as the logical decision element is concerned.

A transistor is a highly complex device whose action is involved and difficult to understand completely. In logical decision elements, however, the treatment may be simplified by regarding it as a switch. Figure 1.22 shows the normal method of representing a transistor diagrammatically and the names of its three connexions.

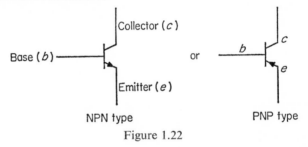

Figure 1.22

Its operation as a switch may be understood by reference to Figure 1.23.

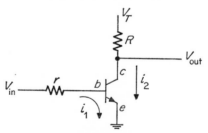

Figure 1.23

If by the application of a suitable positive voltage V_{in} to the base a current i_1 is caused to flow from base to emitter then the transistor will be 'switched on' and i_2 will flow from collector to emitter (provided the voltage V_T is present). With no current i_1, the transistor is 'switched off' and no current i_2 flows.

It is more convenient to discuss the operation of transistor logical decision elements in terms of voltages rather than currents. It is also usual for transistors in this type of application to be operated so that when switched on the resistances from b to e and from c to e are very small. When switched off, however, they are large.

Thus in Figure 1.23 the application of a positive voltage V_{in} to the input will result in an output V_{out} of approximately zero volts. This is because V_{in} causes a current i_1 to flow resulting in current i_2 flowing and the transistor being switched on. The resistance between c and e is therefore much lower than R and V_{out} is almost at the same voltage as e, i.e. zero volts. Conversely, if V_{in} is zero volts the output V_{out} is equal to the positive voltage V_T. These two voltage levels provide our two binary states to which we may assign the truth values T for V_T and F for zero volts.

Note that the basic transistor switch has

an output corresponding to T for an input corresponding to F

and

an output corresponding to F for an input corresponding to T

and behaves as a negation decision element.

Let us now consider the circuit of Figure 1.24. This shows a decision element for the functor $D_k(\ ,\ ...,\ )$ connected to a transistor switch. Thus the output will correspond to the functor $D_k^*(\ ,\ ...,\ )$ where

$$D_k^*(P_1, ..., P_k) =_T \sim D_k(P_1, ..., P_k) \quad (k = 2, 3, ...)$$

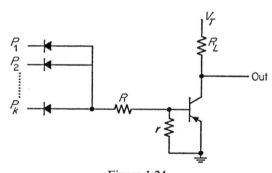

Figure 1.24

If for some integer i $(1 \leqslant i \leqslant k)$ the point corresponding to P_i is at a positive voltage level V_T then the transistor will be switched on and the output V_{out} will be at zero volts. If, on the other hand, all inputs are at zero volts the transistor will be off and V_{out} will be V_T.

A circuit for a joint denial decision element is shown in Figure 1.25.*

This consists essentially of two transistors connected back to back. In this form current can flow through the resistor R whenever one or other or both of the transistors are switched on. Thus if at least one input P, Q is at voltage level V_T then the output is at zero.

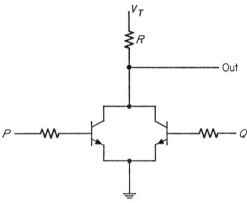

Figure 1.25

In the one remaining case where both P and Q are at zero volts neither transistor conducts and the output is at voltage level $+ V_T$.

Examples 1F

1. Design a transistor circuit for the functor $C_5(\ ,\ ...,\ )$ using the convention that $V_T > 0$. If the same hardware is used with the convention $V_T < 0$, to what functor does it correspond?
2. If $C_k{}^*(P_1, ..., P_k) =_T \sim C_k(P_1, ..., P_k)$ $(k = 2, 3, ...)$ design a transistor circuit whose outputs correspond to the formula $C_3{}^*(P_1, P_2, P_3)$, using the convention $V_T < 0$.

Solutions 1F

1.

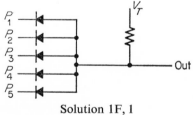

Solution 1F, 1

If $V_T < 0$ we must consider the dual functor, i.e. $D_5(\ ,\ ...,\ )$.

* This circuit was published by Wickes.[10]

1.10 Series parallel circuits. Relationship with the propositional calculus and Boolean algebra

The relay decision elements of Sections 1.3, 1.7 may be constructed using series parallel circuits[1] of relay switches. Let us consider first the case of two switches in series as shown in Figure 1.26.

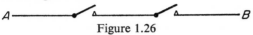

Figure 1.26

In this case, if a potential difference is applied between the points A and B, a current will flow between these points if and only if both switches are closed. Similarly, for switches in parallel, a current will flow between these points if and only if at least one of the two switches is closed (Figure 1.27).

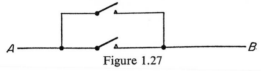

Figure 1.27

Similar considerations apply whenever there are k switches in series or k switches in parallel ($k = 2, 3, \ldots$). Thus, if the switches are operated by relays and we use the same physical representations of truth values as in Section 1.3, the circuits shown in Figures 1.28 and 1.29 will act as decision elements for

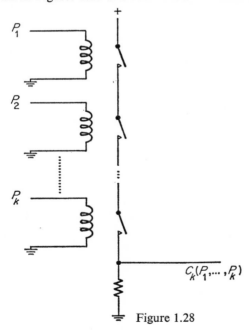

Figure 1.28

the functors $C_k(\,,\ldots,)$, $D_k(\,,\ldots,)$ $(k = 2, 3, \ldots)$. More generally, for any integer n $(n \geqslant 2)$, if the formula $\Phi(P_1, \ldots, P_n)$ is constructed from the formulae $P_1, \ldots, P_n$ using only some of the functors

$$C_2(\,,), \quad C_3(\,,,), \ldots, \quad D_2(\,,), \quad D_3(\,,,), \ldots$$

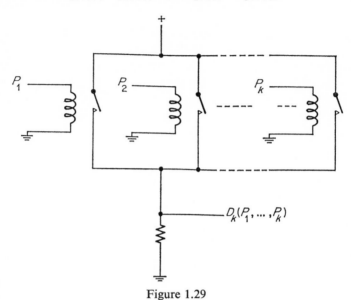

Figure 1.29

and the truth table of the functor $F(\,,\ldots,)$ of n arguments is defined by the equation

$$F(P_1, \ldots, P_n) =_{\mathrm{T}} \Phi(P_1, \ldots, P_n)$$

we may use the above conditions for the existence of closed circuits between A and B to construct a decision element for the functor $F(\,,\ldots,)$.

For example, let us consider the formula

$$[\![(p \vee q \,\&\, r) \vee q \,\&\, s]\!] \,\&\, (r \,\&\, s)$$

If the switches marked p, q, r, s close if and only if their operating relays have their non-earthed coil connexions in the physical states corresponding to the assignment of the truth value T to the variable in question, we see (Figure 1.30) that current can flow from A to B if and only if it can do so via at least one of the points D, E, F. Thus the condition for current to flow from A to B is the (triple) disjunction of the conditions for it to flow via D, E and F. Hence we may take the first of the three disjuncts to be p and the remaining two to be as previously determined for switches in series, i.e. $q \,\&\, r$, $q \,\&\, s$.

We have therefore found that current can flow from A to B (when a potential difference is applied between these points) if and only if the physical conditions represent a case where the formula

$$(p \vee q \,\&\, r) \vee q \,\&\, s$$

takes the truth value T. Similarly current can flow from B to C (when a potential difference is applied between these latter points) if and only if (the physical conditions represent a case where) the formula $r \,\&\, s$ takes the truth value T.

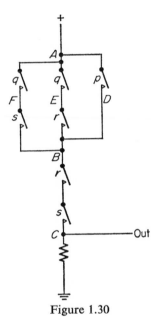

Figure 1.30

Since there will be a closed circuit between A and C if and only if there are closed circuits between A and B and between B and C, the condition for current to flow from A to C when a potential difference is applied between these points is

$$[\![(p \vee q \,\&\, r) \vee q \,\&\, s]\!] \,\&\, (r \,\&\, s)$$

Although seven switches are shown in Figure 1.30, only four relays need to be used, since a relay may operate several switches.

If the two positions of a relay switch are interchanged, so that the switch is closed if and only if the corresponding relay coil is not energized, the formula now corresponding to the circuit will be obtained from the original one by replacing the corresponding occurrence of the relevant propositional variable by its negation. Thus, for example, since

$$P \supset Q =_T \sim P \vee Q$$

we may construct a decision element for implication as shown in Figure 1.31.

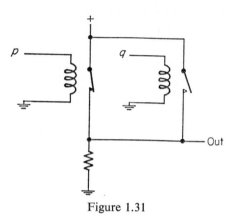

Figure 1.31

Similarly (omitting, as in one previous case, the relay coils) a decision element for the functor whose truth table is determined by the formula

$$[\![(p \& \sim q \vee p \& r) \& (\sim p \& \sim r \vee \sim q \& s)]\!] \& (r \& \sim s)$$

may be constructed as shown in Figure 1.32.

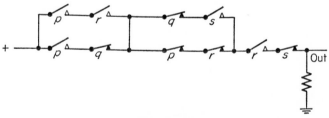

Figure 1.32

It is not necessary, in order to consider the equivalence and non-equivalence functors, to make straightforward uses of equations such as

$$P \equiv Q =_T P \& Q \vee \sim P \& \sim Q \qquad P \not\equiv Q =_T P \& \sim Q \vee \sim P \& Q$$

to provide circuits such as that shown in Figure 1.33 for equivalence. The two switches operated by the relay corresponding to the formula P may be replaced by a two-way switch, as may the two other switches. Thus a decision element for equivalence may be constructed as shown in Figure 1.34 and similarly that for non-equivalence.

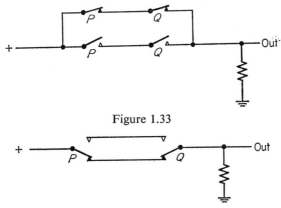

Figure 1.33

Figure 1.34

Examples 1G

1. Draw diagrams for relay circuits corresponding to the formulae
   (i) $[\![(p \& q \lor p \& r) \& s \lor q \& r]\!] \lor p \& s$
   (ii) $\sim p \& \sim q \lor \sim r \& p$
   (iii) $(p/q) \supset (r \equiv s) \lor [\![p \downarrow (q \not\supset r)]\!]$
2. Show how to construct a decision element for non-equivalence, using only two relay-operated two-way switches.
3. Draw diagrams for relay circuits corresponding to the formulae
   (i) $(p \supset q \lor r)/(p \& q \not\equiv s)$
   (ii) $KCpqAqKBprJps$
4. Show how to construct a decision element for conditioned disjunction, using two relay-operated switches and one relay-operated two-way switch.
5. Find, wherever possible, simplifications in circuits considered, similar to those discussed at the end of Solutions 1G.

Solutions 1G

1. (i)

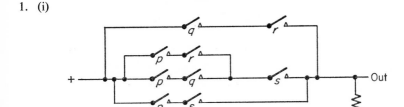

Solution 1G1 (i)

(ii)

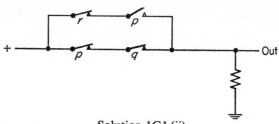

Solution 1G1 (ii)

(iii) $(p/q) \supset (r \equiv s) \vee [\![p \downarrow (q \not\equiv r)]\!]$
$=_T [\![\sim (p/q) \vee (r \equiv s)]\!] \vee \sim p \ \& \ (\sim q \vee r)$
$=_T [\![p \ \& \ q \vee (r \equiv s)]\!] \vee \sim p \ \& \ (\sim q \vee r)$

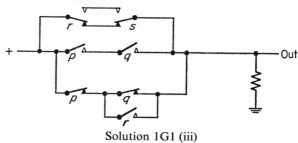

Solution 1G1 (iii)

2.

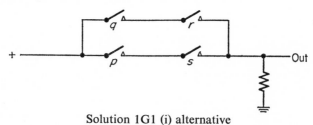

Solution 1G2

We note that, in 1 (i),
$[\![(p \ \& \ q \vee p \ \& \ r) \ \& \ s \vee q \ \& \ r]\!] \vee p \ \& \ s =_T [\![(p \ \& \ q \vee p \ \& \ r) \ \& \ s \vee p \ \& \ s]\!] \vee q \ \& \ r$
$=_T [\![(p \ \& \ q \vee p \ \& \ r) \vee p]\!] \ \& \ s \vee q \ \& \ r$
$=_T p \ \& \ s \vee q \ \& \ r$
giving the simpler circuit shown below.

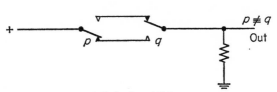

Solution 1G1 (i) alternative

It often happens that electric circuits are represented, not by the propositional calculus directly, but by an algebraic model of this calculus known as Boolean algebra. If the logical operations $\vee \, \& \sim$ considered above are replaced by the operations $+ \, . \, '$ respectively, the propositional variables are replaced by variables of the algebra and $=_T$ is replaced by $=$ then we obtain, from a logical equation, a valid equation of Boolean algebra. For example, the equation

$$\sim(p \, \& \, q \vee r) =_T (\sim p \vee \sim q) \, \& \sim r$$

of the propositional calculus could properly be replaced by the Boolean equation

$$(A \, . \, B + C)' = (A' + B') \, . \, C'$$

Boolean algebra may be developed as an axiomatic system* and it follows easily that, by means of the interpretation considered above, propositions form an example of a Boolean algebra. It can also be easily checked that classes form another such example, the operations $+ \, . \, '$ being regarded as set-theoretical union, intersection and complement respectively. In fact, an equation of the propositional calculus is valid if and only if the corresponding Boolean equation is valid with respect to its set-theoretic interpretation. This follows by considering the correspondence between propositional variables $p, q, r, \ldots$ and statements of the form $x \in A_p, x \in A_q, x \in A_r, \ldots$, since the truth values of these statements may, for general set-theoretical interpretations of the variables $A_p, A_q, A_r, \ldots$, be varied independently. (In the remainder of this section we shall not always show association by brackets.)

For example the equation

$$\sim(p \, \& \, q) =_T \sim p \vee \sim \sim \sim q$$

is valid, though the equation

$$p \vee q \, \& \, r =_T (p \vee q) \, \& \sim \sim r$$

is not. The set-theoretic equation

$$(A \, . \, B)' = A' + B'''$$

is valid, though the set-theoretic equation

$$A + B \, . \, C = (A + B) \, C''$$

is not, as may be seen from Figure 1.35. The second propositional calculus equation fails whenever p, r take the truth values T, F respectively and,

---

* See, for example, Goodstein.[11]

correspondingly, the second set-theoretic equation fails whenever we consider an element of A which is not also an element of C. This may be seen by inspection of Figure 1.35, since the regions $A + B.C$, $(A + B)C''$ are shaded vertically and horizontally respectively.

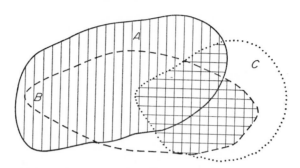

Figure 1.35 A is the region within the continuous curved line. B is the region within the broken line. C is the region within the dotted line

It is easily seen that the value of the set-theoretical expression $A + A'$ (corresponding to the tautology $p \vee \sim p$) is the universal class. Hence a formula of the propositional calculus is a tautology if and only if it is equal to the universal class (since all tautologies are connected by $=_T$). We denote this class by I and its complement, the empty class, by O.

The operations of Boolean addition and multiplication bear a considerable analogy with the elementary operations of the same names. The set-theoretical laws

$$A + B = B + A \qquad\qquad (AB)C = A(BC)$$

$$AB = BA \qquad\qquad A(B + C) = AB + AC$$

$$(A + B) + C = A + (B + C) \qquad A + BC = (A + B)(A + C)$$

follow at once from their propositional calculus counterparts, all six of which have been established in Examples 1A and 1B.

We shall make use of some of these set-theoretical laws in showing algebraically, as an example, that the formula

$$(p \supset q \vee r) \vee \sim \sim p \;\&\; q \vee p \;\&\; (q \supset r)$$

is a tautology. We note first that this formula takes the same truth value as the formula

$$\sim p \vee q \vee r \vee \sim \sim p \;\&\; q \vee p \;\&\; (\sim q \vee r)$$

and then consider the set-theoretical counterpart of this latter formula.

$$A' + B + C + A''B + A(B' + C) = A' + B + C + AB + A(B' + C)$$
$$= A' + B + C + A(B + B' + C)$$
$$= A' + B + C + AI$$
$$= A' + B + C + A$$
$$= A' + A + B + C$$
$$= I + B + C$$
$$= I$$

Examples 1H

1. Rewrite, in the notation of Boolean algebra, the formulae
   (i) $(p \lor q \,\&\, r) \lor \sim (p \,\&\, r)$
   (ii) $(p \supset q) \equiv [\![r \,\&\, \sim (p \lor q)]\!]$
2. Show, algebraically, that the following formulae are tautologies.
   (i) $(p \,\&\, q \lor p \,\&\, \sim q) \lor (q \,\&\, \sim p \lor \sim p \,\&\, \sim q)$
   (ii) $[\![p \supset (q \supset r)]\!] \supset [\![q \supset (\sim r \supset \sim p)]\!]$
3. Prove that a formula is an absurdity if and only if the corresponding set-theoretical expression is equal to O.
4. Classify, by algebraic methods, the following formulae as tautologies, absurdities or mixed formulae.
   (i) $(p \supset q \,\&\, r) \supset (p \supset q) \,\&\, (r \supset p)$
   (ii) $\{(p \supset q) \,\&\, [\![(q \supset r) \,\&\, p]\!]\} \,\&\, \sim r$
   (iii) $(p \lor q \supset r \,\&\, s) \supset [\![(s \supset u) \supset (p \supset u)]\!]$
   (iv) $(p \supset q \lor r) \,\&\, (r \supset s) \supset (p \not\supset s)$
5. Represent, by series parallel relay circuits, the Boolean expressions
   (i) $(A + B)(C + D) + A'C'$
   (ii) $(A + BC)' + D$
   (iii) $[\![(AB + CD)' + E]\!] + B$
   (iv) $[\![(A + B)C + D]\!] + E$

Solutions 1H

1. (i) $(A + BC) + (A + C')'$
   (ii) $(p \supset q) \equiv [\![r \,\&\, \sim (p \lor q)]\!]$
   $\qquad =_\mathrm{T} (\sim p \lor q) \,\&\, r \,\&\, \sim (p \lor q) \lor p \,\&\, \sim q \,\&\, (\sim r \lor p \lor q)$
   which becomes $(A' + B) C(A + B)' + AB'(C' + A + B)$
2. (i) $AB + AB' + BA' + A'B' = AB + AB' + BA' + B'A'$
   $\qquad\qquad\qquad\qquad\qquad\quad = A(B + B') + (B + B') A'$
   $\qquad\qquad\qquad\qquad\qquad\quad = AI + IA'$
   $\qquad\qquad\qquad\qquad\qquad\quad = A + A'$
   $\qquad\qquad\qquad\qquad\qquad\quad = I$
   (ii) $[\![p \supset (q \supset r)]\!] \supset [\![q \supset (\sim r \supset \sim p)]\!] =_\mathrm{T} \sim (\sim p \lor \sim q \lor r) \lor \sim q \lor r \lor \sim p$
   $(A' + B' + C)' + B' + C + A' = (A' + B' + C)' + A' + B' + C$
   $\qquad\qquad\qquad\qquad\qquad\qquad = I$

3. P is an absurdity iff $\sim P$ is a tautology. Thus, if A is the set corresponding to P, P is an absurdity iff $A' = I$, i.e. iff $A = O$.

4. (i) $(A' + BC)' + (A' + B)(C' + A)$

$$= A(B' + C') + (A' + B)(C' + A)$$
$$= [\![A + (A' + B)(C' + A)]\!]\,[\![B' + C' + (A' + B)(C' + A)]\!]$$
$$= (A + A' + B)(A + C' + A)(B' + C' + A' + B)(B' + C' + C' + A)$$
$$= I(A + C')\,I(B' + C' + A)$$
$$= (A + C')(A + C' + B)$$
$$= A + C' \neq I, O \quad \text{(in general)}$$

5. (i)

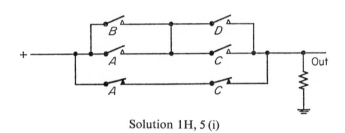

Solution 1H, 5 (i)

(ii) $(A + BC)' + D = A'(B' + C') + D$ (making use of the set-theoretical counter-parts of the de Morgan laws).

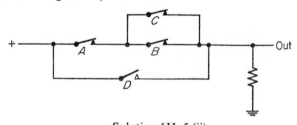

Solution 1H, 5 (ii)

1.11 Applications of logical properties of series parallel circuits to the construction of a control mechanism for the fuelling of a nuclear reactor

This section is an account of some of the work done by the author for the English Electric Company at Whetstone in connexion with their design of a reactor for a nuclear power station. In this reactor, a number of fuel containers are placed on a wheel, rotatable about its axis, the containers being spaced symmetrically. If a container is to be filled the wheel must be rotated so that the container is brought to the top of the wheel. In some cases an anti-clockwise rotation will bring the container to the required position quicker than a clockwise rotation, while in others a clockwise rotation is preferable. Thus, for example, if the wheel has 12 containers and is adjusted as shown in Figure 1.36 with container 9 in the fuelling position and we wish to fill container 1 then we must rotate the wheel anti-clockwise. If, on the other hand, we wish to fill container 6 we must rotate clockwise. If we wish to fill container 3 we must rotate the wheel, but the nature of the decision concerning the direction of rotation is immaterial. If we wish to fill container 9 we do *not* rotate the wheel.

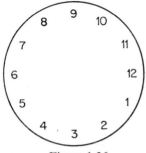

Figure 1.36

The Company wished to use a control mechanism with inputs corresponding to the position of the wheel and the number of the container to be filled and two outputs corresponding respectively to the need to rotate the wheel clockwise or anti-clockwise.

Let us suppose that the total number of containers on the wheel is $2n$ (where n is a positive integer) and that the numbers of the container at present in the fuelling position and of the container which is required to be moved to the fuelling position correspond to the states of two k-tuples of input wires. Each of the $2k$ input wires is capable of two states, which we shall, in turn, make correspond to the truth values T, F respectively. For reasons of safety the Company required that the states of exactly three wires from each of the two k-tuples should correspond to T, the relationship between the states of a set of k wires and the number of a container being otherwise free to be determined by the designer of the control mechanism.

Let the propositional variables $p_1, ..., p_k$ correspond to the k input wires which themselves correspond to the number of the container at present in the fuelling position. Let $q_1, ..., q_k$ correspond, in the same way, to the container which is required to be moved to the fuelling position. Let the formulae P, Q correspond to propositions as follows:

P: the wheel should be rotated anti-clockwise.
Q: the wheel should be rotated clockwise.

The most convenient way to associate the states of a set of input wires (i.e. truth values of $p_1, ..., p_k$) with the number of a fuel container appears to be the following.

3

Let the fuel container have number x $(1 \leqslant x \leqslant 2n)$. Then

(i) If $1 \leqslant x \leqslant n$,

p_1 takes the truth value F.

p_2 takes the truth value T.

If $x_i = 1$ or $x_i = 0$ according as p_i takes the truth value T or the truth value F $(i = 1, ..., k)$ and

$$y = \sum_{i=3}^{k} x_i 2^{k-i} \qquad (1)$$

let the values of y for which exactly two of $x_3, ..., x_k$ are equal to 1 be

$$y_1, ..., y_\alpha$$

where

$$y_1 < y_2 < ... < y_\alpha$$

The truth values of $p_3, ..., p_k$ are those which determine the values of $x_3, ..., x_k$ respectively which correspond (by (1)) to y_x.

(ii) If $n+1 \leqslant x \leqslant 2n$,

p_1 takes the truth value T.

p_2 takes the truth value F.

$p_3, ..., p_k$ take the same truth values as for the number $x - n$.

Thus, for example, if $n = 6$ (as in Figure 1.36), the truth values of $p_1, ..., p_k$ are as shown below*

x	p_1	p_2	p_3	p_4	p_5	p_6	y_x
1	F	T	F	F	T	T	3
2	F	T	F	T	F	T	5
3	F	T	F	T	T	F	6
4	F	T	T	F	F	T	9
5	F	T	T	F	T	F	10
6	F	T	T	T	F	F	12
7	T	F	F	F	T	T	3
8	T	F	F	T	F	T	5
9	T	F	F	T	T	F	6
10	T	F	T	F	F	T	9
11	T	F	T	F	T	F	10
12	T	F	T	T	F	F	12

Let a, b respectively be the numbers of the container in the fuelling position and of the container which is required to be filled. Then we may require† that P takes the truth value T if and only if

either (i) $a < b < a+n$

or (ii) $a \geqslant b+n$

---

* If $k > 6$ the truth values of $p_3, ..., p_{k-4}$ would all be F in each of the 12 cases, so it is sufficient to take $k = 6$.

† Cf. the subsequent remarks concerning the case where $|a - b| = n$.

Similarly we may require* that Q takes the truth value T if and only if
either (iii) $b < a < b+n$
or (iv) $b \geq a+n$
We note that in the cases where

$$|a-b| = n$$

we have arbitrarily laid down that the rotation is anti-clockwise if $a > b$ and clockwise if $a < b$.

We note next that if the numbers $z_1, ..., z_{2n}$ are defined by the equations

$$z_x = y_x + 2^{k-2} \quad (x = 1, ..., n)$$

$$z_x = y_{x-n} + 2^{k-1} \quad (x = n+1, ..., 2n)$$

then

$$z_c > z_d \quad \text{if and only if} \quad c > d \quad (c, d = 1, ..., 2n)$$

and

$$z_{x+n} = z_x + 2^{k-2} \quad (x = 1, ..., n)$$

Thus we may rewrite conditions (i)–(iv) in the respective forms (iR)–(ivR) as follows, where w_x is obtained from z_x by interchanging the first (i.e. most significant) two digits of its binary expansion $(x = 1, ..., 2n)$.

(iR) $z_a < z_b$ and, if $a \leq n, z_b < w_a$
(iiR) $b \leq n$ and not $z_a < w_b$
(iiiR) $z_b < z_a$ and, if $b \leq n, z_a < w_b$
(ivR) $a \leq n$ and not $z_b < w_a$

We note next that p_i corresponds to the proposition:
'The ith digit of z_a is equal to 1' $(i = 1, ..., k)$ and that, if r_i corresponds to the proposition:
'The ith digit of w_a is equal to 1' $(i = 1, ..., k)$ then

$$r_i =_T \sim p_i \quad (i = 1, 2)$$

$$r_i =_T p_i \quad (i = 3, ..., k)$$

Let us now make the definitions

$$L(P_1, Q_1) =_{df} Q_1 \not\supset P_1$$

$$L(P_1, ..., P_{k+1}, Q_1, ..., Q_{k+1})$$

$$=_{df} [L(P_2, ..., P_k, Q_2, ..., Q_k), Q_1 \equiv P_1, Q_1] \quad (k = 1, 2, ...)$$

We note that, for the value of k used in the control mechanism,

$$L(p_1, ..., p_k, q_1, ..., q_k)$$

* Cf. the subsequent remarks concerning the case where $|a-b| = n$.

takes the truth value T if and only if $z_a < z_b$. We then have, by conditions (iR) and (iiR),

$$P =_T L(p_1, ..., p_k, q_1, ..., q_k) \& [\![p_1 \vee L(q_1, ..., q_k, \sim p_1, \sim p_2, p_3, ..., p_k)]\!]$$
$$\vee [\![q_2 \not\supset L(p_1, ..., p_k, \sim q_1, \sim q_2, q_3, ..., q_k)]\!]$$

Similarly, by (iiiR) and (ivR),

$$Q =_T L(q_1, ..., q_k, p_1, ..., p_k) \& [\![q_1 \vee L(p_1, ..., p_k, \sim q_1, \sim q_2, q_3, ..., q_k)]\!]$$
$$\vee [\![p_2 \not\supset L(q_1, ..., q_k, \sim p_1, \sim p_2, p_3, ..., p_k)]\!]$$

Since we are using series parallel circuits it will be convenient to make use of the equations

$$[P, Q, R] =_T P \& Q \vee R \& \sim Q$$
$$P \not\supset Q =_T P \& \sim Q$$
$$[R, P \equiv Q, P] =_T (P \equiv Q) \& R \vee \sim (P \equiv Q) \& P$$
$$=_T (P \equiv Q) \& R \vee P \& \sim Q$$

Thus a circuit corresponding to $L(p_1, ..., p_6, q_1, ..., q_6)$ may be constructed as shown in Figure 1.37, since it follows at once from the last of the above equations that

$$L(p_1, ..., p_{k+1}, p_1, ..., q_{k+1}) =_T (q_1 \equiv p_1) \& L(p_2, ..., p_{k+1}, q_2, ..., q_{k+1}) \vee q_1 \& \sim p_1$$

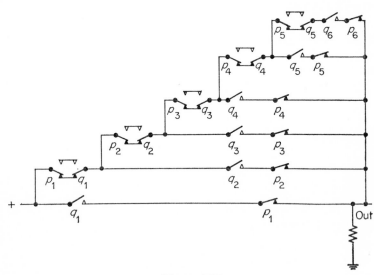

Figure 1.37

Similarly the circuit for P is as shown in Figure 1.39, the circuits for the formulae

$$L(p_1, \ldots, p_6, q_1, \ldots, q_6) \tag{1}$$

$$L(q_1, \ldots, q_6, \sim p_1, \sim p_2, p_3, \ldots, p_6) \tag{2}$$

$$\sim L(p_1, \ldots, p_6, \sim q_1, \sim q_2, q_3, \ldots, q_6) \tag{3}$$

each being represented diagrammatically by a single switch, though, in fact, 22 switches are used in each case. The circuit for

$$L(p_1, \ldots, p_6, \sim q_1, \sim q_2, q_3, \ldots, q_6)$$

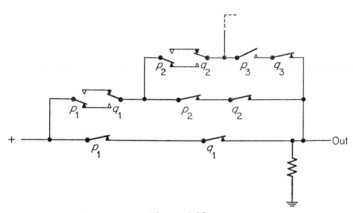

Figure 1.38

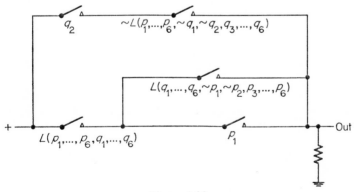

Figure 1.39

is similar to that shown in Figure 1.37, the modifications being shown in Figure 1.38. Similar considerations apply to the circuits for the formulae (2), (3) listed above, a circuit for the formula

$$\sim L(p_1, ..., p_6, \sim q_1, \sim q_2, q_3, ..., q_6)$$

being derived from that of Figure 1.38 by reversing the roles of series and parallel connexions and the positions of all one-way switches when relays are unenergized. The two-way switches remain as shown in Figure 1.38 except that the position of the left-hand switch of each pair is reversed. Thus the circuit for $\sim L(p_1, ..., p_6, \sim q_1, \sim q_2, q_3, ..., q_6)$ is as shown in Figure 1.40. Similar methods will, of course, provide us with a circuit for Q.

Since, in many cases, the energizing of a relay will open a switch instead of closing it the above mechanisms do not 'fail-safe'. However, the control mechanism can be prevented from causing the wrong container to fill by fitting fail-safe mechanisms, such as two relay switches in series, for each formula

$$p_i \& q_i \quad (i = 1, ..., k)$$

and combining these with a 'fail-safe' mechanism for the 'at-least-three' functor, whose output causes the filling of the container to commence. If

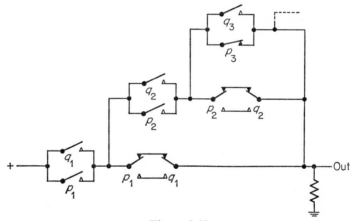

Figure 1.40

$k = 4$ we could use the mechanism shown in Figure 1.41, corresponding to the formula

$$\mathscr{L}_{3,4}(p_1 \& q_1, p_2 \& q_2, p_3 \& q_3, p_4 \& q_4)$$

where $\mathscr{L}_{3,4}(P_1, P_2, P_3, P_4)$ takes the truth value T if and only if at least three of the four formulae P_1, P_2, P_3, P_4 take the truth value T. The eight-variable formula mentioned above cannot take the truth value T unless $z_a = z_b$.

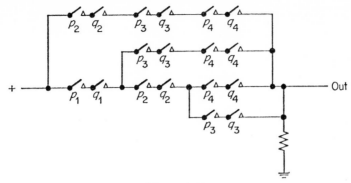

Figure 1.41

1.12 Relay simulation by integrated circuits

Relay switches have been used above to develop logical mechanisms for simplicity. In fact any device which can act as a switch may be used, and it would be more usual to employ transistors, particularly in the form of integrated circuits. It is beyond the scope of this book to deal with the various types of integrated circuits, their construction or limitations.* It is, however, worth noting that relays are very slow, large and expensive, but have complete isolation between the energizing coil and the switch contacts, and may have many separate switches operated by one coil. Transistors, on the other hand, are very fast, minute and inexpensive, but do not have isolation between input and output, and provide one switch only.

---

* For detailed information on this subject see Khambata.[8]

Chapter 2

Definition in the Propositional Calculus, with Practical Applications

2.1 Identities with respect to the relation $=_T$

In various parts of the previous chapter we have seen that formulae of the propositional calculus satisfy, with respect to the relation $=_T$ (though not, of course, with respect to the stronger relation $=$), commutative, associative and distributive laws for conjunction and disjunction and also a number of other laws. For convenience we shall now list these and some further laws.

Commutative law of disjunction	$P \vee Q =_T Q \vee P$
Commutative law of conjunction	$P \& Q =_T Q \& P$
Associative law of disjunction	$(P \vee Q) \vee R =_T P \vee (Q \vee R)$
Associative law of conjunction	$(P \& Q) \& R =_T P \& (Q \& R)$
Idempotent law of disjunction	$P \vee P =_T P$
Idempotent law of conjunction	$P \& P =_T P$
First distributive law	$P \& (Q \vee R) =_T P \& Q \vee P \& R$
Second distributive law	$P \vee Q \& R =_T (P \vee Q) \& (P \vee R)$
Negation cancellation law	$\sim\sim P =_T P$
First de Morgan law	$\sim(P \vee Q) =_T \sim P \& \sim Q$
Second de Morgan law	$\sim(P \& Q) =_T \sim P \vee \sim Q$

Other Laws

$$P \supset Q =_T \sim P \vee Q$$

$$P \not\supset Q =_T P \& \sim Q$$

$$P \equiv Q =_T P \& Q \vee \sim P \& \sim Q =_T (P \supset Q) \& (Q \supset P)$$

$$P \not\equiv Q =_T P \& \sim Q \vee \sim P \& Q =_T (P \vee Q) \& \sim(P \& Q) =_T \sim(P \equiv Q)$$

$$P/Q =_T \sim P \vee \sim Q =_T \sim(P \& Q)$$

$$P \downarrow Q =_T \sim P \& \sim Q =_T \sim(P \vee Q)$$

$$[P, Q, R] =_T P \& Q \vee R \& \sim Q =_T (R \vee Q) \& (P \vee \sim Q) =_T (Q \supset P) \& (\sim Q \supset R)$$

$$\sim P \equiv Q =_T P \not\equiv Q =_T P \equiv \sim Q$$

Those laws which have not been discussed in Chapter 1 should be verified by the reader. We note that, by the de Morgan and negation cancellation laws,

$$P \mathbin{\&} Q =_T \mathord{\sim}\mathord{\sim}(P \mathbin{\&} Q) =_T \mathord{\sim}(\mathord{\sim}P \vee \mathord{\sim}Q)$$

$$P \vee Q =_T \mathord{\sim}\mathord{\sim}(P \vee Q) =_T \mathord{\sim}(\mathord{\sim}P \mathbin{\&} \mathord{\sim}Q)$$

and, by the commutative (disjunction) law, the negation cancellation law, and the equation $P \supset Q =_T \mathord{\sim}P \vee Q$ we see that

$$P \supset Q =_T Q \vee \mathord{\sim}P =_T \mathord{\sim}\mathord{\sim}Q \vee \mathord{\sim}P =_T \mathord{\sim}Q \supset \mathord{\sim}P$$

$$\mathord{\sim}P \supset Q =_T \mathord{\sim}\mathord{\sim}P \vee Q =_T P \vee Q =_T Q \vee P =_T \mathord{\sim}\mathord{\sim}Q \vee P =_T \mathord{\sim}Q \supset P$$

$$P \supset \mathord{\sim}Q =_T \mathord{\sim}P \vee \mathord{\sim}Q =_T \mathord{\sim}Q \vee \mathord{\sim}P =_T Q \supset \mathord{\sim}P$$

Examples 2A

1. Prove that
    (i) $(P \vee Q) \mathbin{\&} R =_T P \mathbin{\&} R \vee Q \mathbin{\&} R$
    (ii) $P \vee (Q \vee R) =_T Q \vee (P \vee R)$
    (iii) $\mathord{\sim}P / \mathord{\sim}Q =_T P \vee Q$
    (iv) $\mathord{\sim}[P, Q, R] =_T [\mathord{\sim}P, Q, \mathord{\sim}R]$
    (v) $(P \equiv Q) \equiv R =_T P \equiv (Q \equiv R)$
2. Prove that
    (i) $(P \vee Q) \mathbin{\&} (R \vee S) =_T (P \mathbin{\&} R \vee P \mathbin{\&} S) \vee (Q \mathbin{\&} R \vee S \mathbin{\&} Q)$
    (ii) $P \supset (Q \supset R) =_T Q \supset (P \supset R)$
    (iii) $(\mathord{\sim}P \downarrow \mathord{\sim}Q) \mathbin{\not\downarrow} \mathord{\sim}R =_T (\mathord{\sim}Q \downarrow \mathord{\sim}R) \mathbin{\not\downarrow} \mathord{\sim}P$
    (iv) $P \vee [Q, R, \mathord{\sim}S] =_T [P \vee Q, R, S \supset P]$
    (v) $P \equiv Q =_T Q \equiv P$
3. If the formulae Q, R each contain the n propositional variables denoted by the syntactical variables $P_1, \dots, P_n$ and no others, with no functors other than E; P_i occurs a_i times in Q, b_i times in R $(i = 1, \dots, n)$ and

$$b_1 = a_1 - 2$$

$$b_i = a_i \quad (i = 2, \dots, n)$$

    prove that

$$Q =_T EEP_1 P_1 R =_T R$$

4. Prove that, with the notation of question 3,
    (i) if $a_1, \dots, a_n$ are not all even then Q is mixed,
    (ii) if $a_1, \dots, a_n$ are all even then Q is a tautology.

Solutions 2A

1. (i) $(P \vee Q) \mathbin{\&} R =_T R \mathbin{\&} (P \vee Q) =_T R \mathbin{\&} P \vee R \mathbin{\&} Q =_T P \mathbin{\&} R \vee Q \mathbin{\&} R$
    (ii) $P \vee (Q \vee R) =_T (P \vee Q) \vee R =_T (Q \vee P) \vee R =_T Q \vee (P \vee R)$
    (iii) $\mathord{\sim}P / \mathord{\sim}Q =_T \mathord{\sim}\mathord{\sim}P \vee \mathord{\sim}\mathord{\sim}Q =_T P \vee Q$

(iv) If Q takes the truth value T then

$$[P, Q, R] =_T P \text{ and}$$

$$\sim [P, Q, R] =_T \sim P =_T [\sim P, Q, \sim R]$$

The other case is similar.

(v) If Q takes the truth value T then

$$(P \equiv Q) \equiv R =_T P \equiv R =_T P \equiv (Q \equiv R)$$

If Q takes the truth value F then

$$(P \equiv Q) \equiv R =_T \sim P \equiv R =_T P \equiv \sim R =_T P \equiv (Q \equiv R)$$

3. P_i occurs in $EEP_1 P_1 R$ a_i times ($i = 1, ..., n$). Hence, by Solutions 2A, 1 (v) and Examples 2A, 2 (v),

$$Q =_T EEP_1 P_1 R$$

Since $EP_1 P_1$ is a tautology,

$$EEP_1 P_1 R =_T R$$

4. (i) Let $c_i = 1$ when a_i is odd and $c_i = 0$ when a_i is even ($i = 1, ..., n$). Thus there exists at least one integer i ($1 \leqslant i \leqslant n$) such that $c_i = 1$. By repeated use of the final result of the previous question,

$$Q =_T EP_{i_1} EP_{i_2} ... EP_{i_{k-1}} P_{i_k}$$

where $i_1, ..., i_k$ are the values of i for which $c_i = 1$. (We note that the solution to question 3 remains valid if $b_1 = 0$.) Since the formula

$$
\begin{array}{cccc}
EP_{i_1} & EP_{i_2} & ... \; EP_{i_{k-1}} & P_{i_k} \\
TT & TT & TT & T \\
FF & TT & TT & T
\end{array}
$$

is mixed, so is Q. (If $k = 1$ then $Q =_T P_{i_1}$ and P_{i_1} is mixed.)

2.2 Economy in the use of decision elements. Applications of identities

We saw in the previous chapter that a decision mechanism for any formula of the propositional calculus could be constructed from decision elements for the functors occurring in that formula. It is not, however, necessary to have available a supply of decision elements for every functor. For example, since

$$P \supset Q =_T \sim P \vee Q$$

a decision element for implication may be simulated by decision elements for negation and disjunction. Whatever formulae are denoted by P, Q, the output of a decision mechanism for the formula $\sim P \vee Q$ will be the same as the output of a decision element for $P \supset Q$. (See Figure 2.1.)

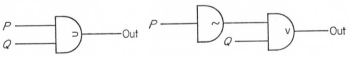

Figure 2.1

More generally, if $F(,...,)$ is a functor of n $(n \geqslant 1)$ arguments and

$$F(P_1, ..., P_n) =_T Q$$

where Q is a formula built up exclusively from sub-formulae $P_1, ..., P_n$ by means of k functors then a decision element for the functor $F(,...,)$ may be simulated by decision elements for these k functors. If the formula Q contains the logical constants t, f (denoting respectively a true or false statement) these may be simulated by keeping the corresponding input wires permanently in the two related physical states (cf. Examples 2B, 3).

Examples 2B

1. Show how to construct a decision mechanism for the formula $(p \supset q) \lor p \& r$ using only decision elements for disjunction and negation. Draw a block diagram.
2. Show how to construct a decision mechanism for the formula $(p \equiv q) \lor (p \supset r)$ using only decision elements for incompatability.
3. Show how to construct a decision mechanism for the formula $p \& q \lor r$ using only decision elements for the ternary functor $F(,,)$, where

$$F(P, Q, R) =_T (P \lor Q) \& R$$

4. Repeat question 3 for implication decision elements.

Solutions 2B

1. $\qquad (p \supset q) \lor p \& r =_T (\sim p \lor q) \lor \sim (\sim p \lor \sim r)$

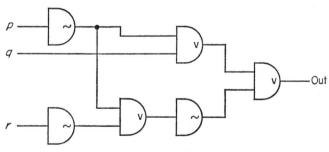

Solution 2B, 1

3. $\qquad P \& Q =_T F(f, P, Q)$

$\qquad\qquad P \lor Q =_T F(P, Q, t)$

Hence

$\qquad p \& q \lor r =_T F(p \& q, r, t)$

$\qquad\qquad =_T F(F(f, p, q), r, t)$

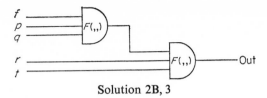

Solution 2B, 3

(The extreme top and bottom wires are kept permanently in the second and first physical states respectively.)

2.3 Alternative systems of 2-valued propositional calculi. The concept of primitive functors and its physical interpretation

We saw in the last chapter that, using only the disjunction and negation functors, we could construct a formula from the sub-formulae P, Q in such a way that it always took the same truth value as $P \supset Q$. We can make similar constructions for the other (non-trivial) binary functors since

$$P \& Q =_\text{T} \sim(\sim P \vee \sim Q)$$

$$P \not\supset Q =_\text{T} \sim(P \supset Q) =_\text{T} \sim(\sim P \vee Q)$$

$$P \equiv Q =_\text{T} P \& Q \vee \sim P \& \sim Q =_\text{T} \sim(\sim P \vee \sim Q) \vee \sim(P \vee Q)$$

$$P \not\equiv Q =_\text{T} (P \vee Q) \& \sim(P \& Q) =_\text{T} (P \vee Q) \& (\sim P \vee \sim Q)$$

$$P/Q =_\text{T} \sim P \vee \sim Q$$

$$P \downarrow Q =_\text{T} \sim(P \vee Q)$$

Thus, instead of considering a propositional calculus with (among other things) 8 different binary functors, we may consider a propositional calculus which uses no functor symbols other than $\vee$ and $\sim$. These are known as the *primitive functors* since they are undefined *in* the propositional calculus, though in talking *about* the calculus we consider the truth table *interpretations* of these functors. The other 7 binary functors may be reintroduced by *definitions* such as

$$P \& Q =_\text{df} \sim(\sim P \vee \sim Q)$$

Thus if, for some formulae P, Q we write $P \& Q$ we have made an *abbreviation* for the formula $\sim(\sim P \vee \sim Q)$. It is reasonable to use the symbol &, which is no longer a symbol of the calculus proper, as an abbreviation in this way since the formula which we have now agreed to abbreviate by $P \& Q$ always takes the same truth value as the formula $P \& Q$ considered in the past. It is not, of course, possible to prove a definition, but since, in the propositional calculus of Chapter 1,

$$P \& Q =_\text{T} \sim(\sim P \vee \sim Q)$$

our present convention of abbreviating, in this propositional calculus with only two primitive functors, the formula $\sim(\sim P \vee \sim Q)$ by $P \& Q$ is reasonable.

The abbreviation applies, of course, to *all* formulae P, Q. Thus, for example, if we make the corresponding definition of equivalence, i.e.

$$P \equiv Q =_{\mathrm{df}} \sim(\sim P \vee \sim Q) \vee \sim(P \vee Q)$$

then the formula

$$p \& q \equiv r \& \sim q$$

is an abbreviation for

$$\sim[\![\sim(p \& q) \vee \sim(r \& \sim q)]\!] \vee \sim(p \& q \vee r \& \sim q)$$

and this, in turn, is an abbreviation for

$$\sim[\![\sim \sim(\sim p \vee \sim q) \vee \sim \sim(\sim r \vee \sim \sim q)]\!] \vee \sim[\![\sim(\sim p \vee \sim q) \vee \sim(\sim r \vee \sim \sim q)]\!]$$

We could, as an alternative, have made the definition

$$P \& Q =_{\mathrm{df}} \sim(\sim Q \vee \sim \sim \sim P)$$

and this is a reasonable definition, since, in the propositional calculus of Chapter 1,

$$P \& Q =_{\mathrm{T}} \sim(\sim Q \vee \sim \sim \sim P)$$

However, it is essential, when describing a propositional calculus, to make it quite clear which abbreviation convention is to be used so that formulae written with defined functors are not ambiguous. For instance, if it is not made absolutely clear which of the definitions

$$P \& Q =_{\mathrm{df}} \sim(\sim P \vee \sim Q)$$

$$P \& Q =_{\mathrm{df}} \sim(\sim Q \vee \sim \sim \sim P)$$

is being used, it would not be clear whether the formula

$$r \& (p \vee q)$$

was an abbreviation for the formula

$$\sim[\![\sim r \vee \sim(p \vee q)]\!]$$

or for the formula

$$\sim[\![\sim(p \vee q) \vee \sim \sim \sim r]\!]$$

Clearly, to every formula of a propositional calculus with given primitive functors there corresponds a decision mechanism constructed entirely from decision elements for these primitive functors. If another functor is (reasonably) definable in this propositional calculus then a decision element for this functor may be simulated by a decision mechanism for the defining formula (apart from decision elements corresponding to functors in the formulae

denoted by syntactical variables). For example, if the primitive functors are
∨ and ∼ then, since we may make the definition

$$P \equiv Q =_{\mathrm{df}} \sim(\sim P \vee \sim Q) \vee \sim(P \vee Q)$$

a decision element for equivalence may be simulated as shown in Figure 2.2.

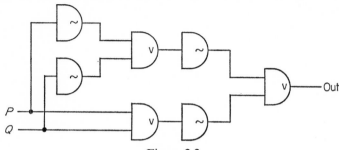

Figure 2.2

The primitive functors may include logical constants (which we may regard
as functors of degree 0). For example, if the primitives are C and f, we may
(reasonably) make the definitions

$$NP =_{\mathrm{df}} CPf$$

$$APQ =_{\mathrm{df}} CNPQ \quad \text{(which abbreviates } CCPfQ)$$

$$KPQ =_{\mathrm{df}} NCPNQ \quad \text{(which abbreviates } CCPCQff)$$

(cf. Solutions 2B, 3).

Examples 2C

1. In the propositional calculus with conjunction and negation as the primitive
   functors define the other seven non-trivial binary functors.
2. Prove that, in the propositional calculus with conjunction as the only primitive
   functor, there is no reasonable definition of negation and vice versa.
3. Establish results corresponding to those of questions 1 and 2 for the propositional calculus whose primitive functors are non-implication and equivalence.

Solutions 2C

1. $P \vee Q =_{\mathrm{df}} \sim(\sim P \& \sim Q)$

   $P \not\supset Q =_{\mathrm{df}} P \& \sim Q$

   $P \supset Q =_{\mathrm{df}} \sim(P \not\supset Q) \quad \text{(which abbreviates } \sim(P \& \sim Q))$

   $P \equiv Q =_{\mathrm{df}} (P \supset Q) \& (Q \supset P) \quad \text{(see above)}$

   $P \not\equiv Q =_{\mathrm{df}} \sim(P \equiv Q) \quad \text{(see above)}$

   $P/Q =_{\mathrm{df}} \sim(P \& Q)$

   $P \downarrow Q =_{\mathrm{df}} \sim P \& \sim Q$

2. Let the syntactical variables $P_1, \ldots, P_n$ denote propositional variables and let $\Phi(P_1, \ldots, P_n)$ be a formula of length l containing no propositional variables other than those denoted by $P_1, \ldots, P_n$ and no functors other than conjunction. We shall prove, by strong induction on l, that, if $P_1, \ldots, P_n$ take the truth value F, so does $\Phi(P_1, \ldots, P_n)$.

If $l = 1$ then $\Phi(P_1, \ldots, P_n)$ is, for some integer i $(1 \leqslant i \leqslant n)$, P_i and the result becomes 'if $P_1, \ldots, P_n$ take the truth value F then P_i takes the truth value F', and this is trivial. We now assume the result for all positive integers less than l and deduce it for l.

Since we have disposed of the case $l = 1$ we may assume that, from now on, $l \geqslant 2$. Thus $\Phi(P_1, \ldots, P_n)$ will be of the form

$$\Psi(P_1, \ldots, P_n) \,\&\, \Lambda(P_1, \ldots, P_n)$$

where no propositional variables other than (those denoted by) $P_1, \ldots, P_n$ occur in $\Psi(P_1, \ldots, P_n)$. Thus, since

$$l(\Phi) = l(\Psi) + l(\Lambda) + 1$$
$$l(\Psi) < l(\Phi)$$

and the induction hypothesis applies to $\Psi(P_1, \ldots, P_n)$. Hence, as easily shown below, when $P_1, \ldots, P_n$ take the truth value F, $\Psi(P_1, \ldots, P_n)$ takes the truth value F, as does $\Phi(P_1, \ldots, P_n)$.

$$\Psi(P_1, \ldots, P_n) \,\&\, \Lambda(P_1, \ldots, P_n) \; [\![= \; \Phi(P_1, \ldots, P_n)]\!]$$
$$F \;\; F \;\; \ldots \; F \quad\;\; F$$

If we may reasonably make the definition

$$\sim P =_{\mathrm{df}} \Phi(P) \tag{A}$$

then, in the propositional calculus of Chapter 1,

$$\sim P =_{\mathrm{T}} \Phi(P)$$

for all formulae P and, in particular,

$$\sim p =_{\mathrm{T}} \Phi(p) \tag{1}$$

But, taking $n = 1$ and p as an instance of P_1, $\Phi(p)$ takes the truth value F when p takes the truth value F, contradicting equation (1). Thus our hypothesis that we may reasonably make the definition (A) is incorrect.

In the propositional calculus with negation as the only primitive functor, no formula contains more than one propositional variable. Hence conjunction is not definable.

2.4 Functional completeness of propositional calculi, with practical applications

In Section 2.3 we saw that, in the propositional calculus with $\lor$ and $\sim$ as primitive functors, all non-trivial binary functors were (reasonably) definable. This result leads us to enquire whether, given an arbitrary positive integer n and an arbitrary truth table of degree n, we can, in this propositional calculus, always construct a formula having that truth table. In fact we can do so, all 2^{2^n} formulae being constructable for each positive integer n. We therefore say that the propositional calculus with $\vee$ and $\sim$ as the primitive functors is *functionally complete*.

We shall prove this by showing, by induction on n, that, for general formulae $P_1, ..., P_n$, we can construct a formula $\Phi(P_1, ..., P_n)$ corresponding to a given truth table of degree n. If $n = 1$ then $2^{2^n} = 4$ and we shall construct formulae $\Phi_i(P)$ ($i = 1, 2, 3, 4$) corresponding to the four truth tables (the suffix of P_1 being omitted as it is unnecessary in this case).

P	$\Phi_1(P)$	$\Phi_2(P)$	$\Phi_3(P)$	$\Phi_4(P)$
T	T	F	T	F
F	F	T	T	F

We may (though there are other suitable choices) take

$$\Phi_1(P) = P$$

$$\Phi_2(P) = \sim P$$

$$\Phi_3(P) = P \vee \sim P$$

$$\Phi_4(P) = \sim(P \vee \sim P)$$

We now assume the result for n and deduce it for $n+1$. If we arbitrarily assign to P_{n+1} the truth value T then the entry in the 'answer' part of our $(n+1)$th degree truth table will be determined entirely by the truth values of $P_1, ..., P_n$. Thus this entry is determined by a truth table of degree n corresponding to which we may, by our induction hypothesis, construct a formula $\Psi(P_1, ..., P_n)$. Similarly we may construct a formula $\Lambda(P_1, ..., P_n)$ which, whenever P_{n+1} takes the truth value F, takes the same truth value as $\Phi(P_1, ..., P_{n+1})$ is required to take. Hence, since we may define conjunction and conditioned disjunction by

$$P \& Q =_{df} \sim(\sim P \vee \sim Q)$$

$$[P, Q, R] =_{df} P \& Q \vee R \& \sim Q$$

we may take $\Phi(P_1, ..., P_{n+1})$ to be the formula

$$[\Psi(P_1, ..., P_n), P_{n+1}, \Lambda(P_1, ..., P_n)]$$

Having established the functional completeness of one propositional calculus, we shall find it relatively easy to establish the functional completeness of other propositional calculi. For instance, in the propositional calculus with $\&$ and $\sim$ as the primitive functors, we may make the definition

$$P \vee Q =_{df} \sim(\sim P \& \sim Q)$$

and the new functional completeness result now follows from the old one. (To construct $\Phi(P_1, ..., P_n)$ we may imitate the old method, using $\sim$ and the *defined* functor $\vee$.)

The propositional calculus with $/$ as the only primitive functor is functionally complete also, since we may make the definitions

$$\sim P =_{\mathrm{df}} P/P$$

$$P \vee Q =_{\mathrm{df}} \sim P / \sim Q$$

If we have a sufficient supply of decision elements for all primitive functors of a functionally complete system (e.g. a supply of disjunction and negation elements) we may construct a decision mechanism for any formula of any propositional calculus, since we may, in the functionally complete calculus, define all functors occurring in the formula. For example, if we are supplied with decision elements for $\vee$ and $\sim$ we may construct a decision mechanism for the formula

$$[p, q, \sim r] \supset [q, \sim p, p/s]$$

since we may (reasonably) make the definitions

$$P \& Q =_{\mathrm{df}} \sim (\sim P \vee \sim Q)$$

$$[P, Q, R] =_{\mathrm{df}} P \& Q \vee R \& \sim Q$$

$$P \supset Q =_{\mathrm{df}} \sim P \vee Q$$

$$P / Q =_{\mathrm{df}} \sim P \vee \sim Q$$

Examples 2D

1. Prove that the propositional calculus with conditioned disjunction and negation as primitive functors is functionally complete and that the primitive functors are independent (cf. Solutions 2C, 2).
2. Prove that the propositional calculus with $\downarrow$ as the only primitive functor is functionally complete. Prove also that this functor cannot be replaced by any binary functor other than $/$.
3. Draw a block diagram, using only $\vee$ and $\sim$ decision elements, of a decision mechanism for the formula

$$[p, q, \sim r] \supset [q, \sim p, p/s]$$

Solutions 2D

1. $P \vee Q =_{\mathrm{df}} [P, P, Q]$

$$[P, Q, R]$$
$$TT \ \ T$$

$$\Phi(P_1, ..., P_n) \quad \text{(no occurrences of } \sim \text{)}$$
$$T \ \ T \quad \ \ T$$

2.5 Sheffer functions and pseudo-Sheffer functions of machine logic

In Section 2.4 it was proved that any decision mechanism could be constructed from decision elements for functors determining a functionally complete system. Thus, in particular, it is sufficient to have an adequate supply of decision elements for the functor $/$ or for the functor $\downarrow$ (cf. Examples 2D, 2). The requirement of functional completeness is, however, too restrictive, since the logical constants t, f may always be simulated (cf. Solutions 2B, 3). It is therefore sufficient to have a supply of decision elements for a functor $F(,\dots,)$ such that the propositional calculus with $F(,\dots,)$, t and f as primitives is functionally complete.

We say that a binary functor $F(,)$ is a *Sheffer function* if the propositional calculus with $F(,)$ as the only primitive is functionally complete (so that $F(,)$ must be $/$ or $\downarrow$ in the 2-valued case) and that $F(,)$ is a *pseudo-Sheffer function of machine logic* if

(i) $F(,)$ is not a Sheffer function,

(ii) the propositional calculus with $F(,)$, t and f as primitives is functionally complete.

For example, implication is a pseudo-Sheffer function of machine logic, since the propositional calculus with implication as the only primitive functor is not functionally complete (cf. Examples 2D, 2) and we may, in the propositional calculus whose primitives are implication and the logical constant f, make the definitions

$$\sim P =_{\mathrm{df}} P \supset f$$

$$P \vee Q =_{\mathrm{df}} \sim P \supset Q$$

$$t =_{\mathrm{df}} f \supset f$$

(We do not regard systems without logical constants as functionally incomplete, but, if one constant is a primitive, we require the other to be definable for functional completeness.)

Examples 2E

1. Prove that non-implication is a pseudo-Sheffer function of machine logic and show how to simulate a conjunction decision element, using only non-implication decision elements.
2. Prove that if

$$\Phi(P, Q, R, S) =_{\mathrm{T}} P \equiv [\![Q \equiv (R \supset S)]\!]$$

then a single decision element for the functor $\Phi(,,,)$ may simulate a decision element for any of the non-trivial binary functors.
3. Prove that disjunction is not a pseudo-Sheffer function of machine logic.

Solutions 2E

1.
$$\sim P =_{df} t \not\supset P$$
$$P \supset Q =_{df} \sim(P \not\supset Q)$$
$$P \lor Q =_{df} \sim P \supset Q$$
$$P \,\&\, Q =_{df} P \not\supset \sim Q \quad \text{(which abbreviates } P \not\supset(t \not\supset Q)\text{)}$$

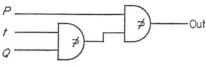

Solution 2*E*, 1

2.
$$P \lor Q =_{df} \Phi(t, P, Q, P)$$

Solution 2E, 2

Similarly in the other seven cases.

2.6 The double-line technique, an alternative approach

It is sometimes considered desirable, in order to avoid inversions, to use pairs of inputs corresponding to propositional variables[12] instead of single inputs. With each sub-formula P of a given formula, we may associate formulae P_1, P_2 such that

$$P_1 =_T P \qquad P_2 =_T \sim P$$

In particular, if P is a propositional variable occurring in the given formula, there will correspond to P two propositional variables P_1, P_2 the inputs corresponding to which are in the same and opposite physical states respectively as the (single) input which previously corresponded to P.

If $F(\,,)$ is a binary functor we can always find binary functors $F_1(\,,), F_2(\,,)$ such that

$$F_1(X, Y) =_T F(P, Q)$$
$$F_2(Z, W) =_T \sim F(P, Q)$$

for some formulae X, Y, Z, W such that

$$X, Y, Z, W \in \{P_1, P_2, Q_1, Q_2\}$$

and we can, except in two cases, even do this in such a way that

$$F_1(\,,), F_2(\,,) \in \{\lor, \&\}$$

thereby avoiding inversion.

We may determine $F_1(\ ,\ )$, $F_2(\ ,\ )$, X, Y, Z, W, as follows:

$$P \vee Q =_{\mathrm{T}} P_1 \vee Q_1$$

so that we may take X, Y, $F_1(\ ,\ )$ as $P_1, Q_1, \vee$, respectively and we therefore take

$$(P \vee Q)_1 = P_1 \vee Q_1.$$

$$\sim(P \vee Q) =_{\mathrm{T}} \sim P\ \&\sim Q =_{\mathrm{T}} P_2\ \&\ Q_2$$

so that we may take Z, W, $F_2(\ ,\ )$ as $P_2, Q_2, \&$, respectively and we therefore take

$$(P \vee Q)_2 = P_2\ \&\ Q_2$$

Similarly we can show that we may take

$$(P\ \&\ Q)_1 = P_1\ \&\ Q_1 \qquad\qquad (P\ \&\ Q)_2 = P_2 \vee Q_2$$

$$(P \supset Q)_1 = P_2 \vee Q_1 \qquad\qquad (P \supset Q)_2 = P_1\ \&\ Q_2$$

$$(P \not\supset Q)_1 = P_1\ \&\ Q_2 \qquad\qquad (P \not\supset Q)_2 = P_2 \vee Q_1$$

$$(P \equiv Q)_1 = P_1 \equiv Q_1 \qquad\qquad (P \equiv Q)_2 = P_2 \equiv Q_1$$

$$(P \not\equiv Q)_1 = P_2 \equiv Q_1 \qquad\qquad (P \not\equiv Q)_2 = P_1 \equiv Q_1$$

$$(P/Q)_1 = P_2 \vee Q_2 \qquad\qquad (P/Q)_2 = P_1\ \&\ Q_1$$

$$(P \downarrow Q)_1 = P_2\ \&\ Q_2 \qquad\qquad (P \downarrow Q)_2 = P_1 \vee Q_1$$

In the exceptional cases of equivalence and non-equivalence we may, if we wish to use more decision elements rather than use inversion, alternatively take

$$(P \equiv Q)_1 = P_1\ \&\ Q_1 \vee P_2\ \&\ Q_2 \qquad (P \equiv Q)_2 = P_1\ \&\ Q_2 \vee P_2\ \&\ Q_1$$

$$(P \not\equiv Q)_i = (P \equiv Q)_{3-i} \quad (i = 1, 2)$$

As an example of the use of the double-line technique we shall consider now the construction of a decision mechanism for the formula

$$\langle p\ \&\ q \supset [\![p \vee r \equiv (q \downarrow s)]\!] \rangle\ \&\sim r$$

We note that we may take

$$(\sim P)_1 = P_2 \quad (\sim P)_2 = P_1$$

so that, in particular

$$(\sim r)_1 = r_2 \quad (\sim r)_2 = r_1$$

and we recall the equations

$$(P\ \&\ Q)_1 = P_1\ \&\ Q_1 \quad (P\ \&\ Q)_2 = P_2 \vee Q_2$$

together with the corresponding equations for the functors

$$\supset \quad \vee \quad \equiv \quad \downarrow$$

Thus we may use the mechanism illustrated in Figure 2.3. We note that

$$\{\langle p \,\&\, q \supset [\![p \vee r \equiv (q \downarrow s)]\!] \rangle \,\&\, {\sim} r\}_1 = \{ p \,\&\, q \supset [\![p \vee r \equiv (q \downarrow s)]\!] \}_1 \,\&\, ({\sim} r)_1$$

$$\{\langle p \,\&\, q \supset [\![p \vee r \equiv (q \downarrow s)]\!] \rangle \,\&\, {\sim} r\}_2 = \{ p \,\&\, q \supset [\![p \vee r \equiv (q \downarrow s)]\!] \}_2 \vee ({\sim} r)_2$$

replace $({\sim} r)_1$ by r_2, $({\sim} r)_2$ by r_1 and then repeat the process on shorter sub-formulae.

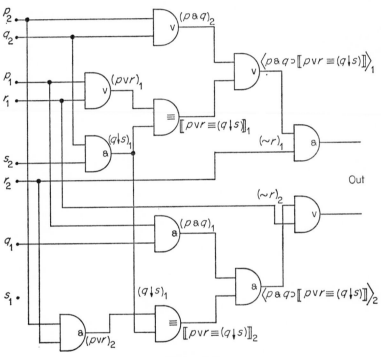

Figure 2.3

Examples 2F

1. Find all formulae $(P \vee Q)_1$, $(P \vee Q)_2$ containing only one binary functor.
2. Show how to construct a double-line decision mechanism for the formula

$$p \,\&\, q \supset {\sim}(p \vee r/q)$$

   Draw a block diagram.
3. Find all formulae $(P \supset Q)_1, (P \supset Q)_2, (P \equiv Q)_1, (P \equiv Q)_2$ containing only one binary functor.

4. Find all inversion-free formulae $(P \equiv Q)_1$, $(P \equiv Q)_2$ containing only three binary functors.

5. Show how to construct an inversion-free mechanism for the formula

$$p \vee q \equiv (p \not\equiv r)$$

Draw a block diagram.

Solutions 2F

1. $(P \vee Q)_1: P_1 \vee Q_1, P_2 \supset Q_1, Q_2 \supset P_1, P_2/Q_2$

   $(P \vee Q)_2: P_2 \& Q_2, P_2 \not\supset Q_1, Q_2 \not\supset P_1, P_1 \downarrow Q_1$

(Formulae such as $Q_1 \vee P_1$, $Q_1 \subset P_2$ are omitted.)

2. We may take

$$\llbracket p \& q \supset \sim (p \vee r/q) \rrbracket_1 = (p \& q)_2 \vee \llbracket \sim (p \vee r/q) \rrbracket_1$$

$$= (p_2 \vee q_2) \vee (p \vee r/q)_2$$

$$= (p_2 \vee q_2) \vee (p \vee r)_1 \& q_1$$

$$= (p_2 \vee q_2) \vee (p_1 \vee r_1) \& q_1$$

$$=_T D_3(p_2, q_2, \llbracket p_1 \vee r_1 \rrbracket \& q_1)$$

$$\llbracket p \& q \supset \sim (p \vee r/q) \rrbracket_2 = (p \& q)_1 \& \llbracket \sim (p \vee r/q) \rrbracket_2$$

$$= (p_1 \& q_1) \& (p \vee r/q)_1$$

$$= (p_1 \& q_1) \& \llbracket (p \vee r)_2 \vee q_2 \rrbracket$$

$$=_T C_3(p_1, q_1, p_2 \& r_2 \vee q_2)$$

Solution 2F, 2

2.7 Reductions to normal forms and their applications (elementary theory)

It is well known that, in elementary algebra, expressions formed from variables and the operations of addition and multiplication can be reduced to sums of products and that, in somewhat more advanced algebra, quadratic forms can be reduced, by complex linear transformations, to sums of squares. In a similar way we shall make use of the equations

$$\sim \sim P =_{\mathrm{T}} P \tag{1}$$

$$\sim (P \vee Q) =_{\mathrm{T}} \sim P \,\&\sim Q \tag{2}$$

$$\sim (P \,\& \,Q) =_{\mathrm{T}} \sim P \vee \sim Q \tag{3}$$

$$P \vee Q \,\& \,R =_{\mathrm{T}} (P \vee Q) \,\& \,(P \vee R) \tag{4A}$$

$$P \,\& \,(Q \vee R) =_{\mathrm{T}} P \,\& \,Q \vee P \,\& \,R \tag{4B}$$

$$P \vee Q =_{\mathrm{T}} Q \vee P \tag{5A}$$

$$P \,\& \,Q =_{\mathrm{T}} Q \,\& \,P \tag{5B}$$

to carry out reductions, with respect to the relation $=_{\mathrm{T}}$, of formulae of the propositional calculus.

A formula P is said to be in *conjunctive normal form* if it satisfies the following conditions.

(i) It contains no functors other than conjunction, disjunction and negation (possibly used as definite unambiguous abbreviations).

(ii) In building up the formula from propositional variables and the three functors, all combination by negation is completed before any combination by conjunction or disjunction is undertaken, and the only sub-formulae of P which are of the form $\sim Q$ are those for which Q denotes a propositional variable (so that formulae such as $\sim \sim p$, $\sim (p \vee q)$, $\sim (p \,\& \,q)$ are not in conjunctive normal form).

(iii) In building up the formula from propositional variables and the three functors, no combination by conjunction is undertaken until all combinations by other functors have been completed.

Thus, for example, the formula

$$\{\langle (p \vee q) \vee [\![(\sim p \vee r) \vee q]\!] \rangle \,\& \,(q \vee \sim r)\} \,\& \,q$$

is in conjunctive normal form, but the formulae

$$(\sim \sim p \vee q) \,\& \,(\sim q \vee r) \qquad\qquad [\![\sim (p \vee q) \vee r]\!] \,\& \sim q$$

$$[\![p \vee \sim (q \vee p)]\!] \,\& \,(r \vee \sim q) \qquad\qquad (p \vee q \,\& \sim r) \,\& \,(\sim r \vee \sim q)$$

are not in conjunctive normal form.

A formula is said to be in *disjunctive normal form* if it satisfies (i) and (ii) and the condition obtained from (iii) by interchanging the roles of conjunction and disjunction. Thus, for example, the formula

$$[\![(p \,\& \sim q) \,\& \sim p \vee (r \,\& \sim q) \,\& \,(p \,\& \,s)]\!] \vee [\![p \,\& \,(\sim s \,\& \,q)]\!]$$

is in disjunctive normal form, but the formula

$$\sim\,\sim p\,\&\sim(q\,\&\,r)\vee\sim p\,\&\,(r\vee p)$$

is not in disjunctive normal form. The formula

$$[\![(p\,\&\sim q)\,\&\,(r\,\&\,q)]\!]\,\&\sim s$$

is in both conjunctive and disjunctive normal forms as is the formula

$$[\![\sim p\vee(q\vee r)]\!]\vee(p\vee\sim s)$$

We shall now show that to every formula P there corresponds a formula P^N such that P^N is in conjunctive normal form and

$$P^N =_T P$$

In a similar way we can find a formula P^n such that P^n is in disjunctive normal form and

$$P^n =_T P$$

The method for obtaining P^n is identical with that for obtaining P^N except equations (4B), (5B) are used in place of equations (4A), (5A) respectively.

In order to carry out the reduction we ensure compliance with conditions (i), (ii), (iii) by proceeding in the following three stages.

(a) Use identities of Section 2.1 (such as $P\supset Q =_T \sim P\vee Q$) to remove functors other than the three permitted in (i).

(b) Use (2) and (3) repeatedly, at every stage making use of (1) as soon as possible, until (ii) is complied with in every respect.

(c) Use (4A) or (4B) repeatedly, together with the corresponding right-hand distributive law (obtained from (4A) and (5A) or from (4B) and (5B)), until (iii) is complied with in every respect.

Let us, for example, consider the reduction of the formula

$$(p\supset q\vee r)\supset[\![p/(r\,\overline{\supset}\,q)]\!]\vee(p\downarrow r)$$

Proceeding in the way indicated above for the conjunctive case we have (omitting brackets showing associations of associative functors),

$(p\supset q\vee r)\supset[\![p/(r\,\overline{\supset}\,q)]\!]\vee(p\downarrow r)$

$=_T \ \sim(\sim p\vee q\vee r)\vee\sim p\vee\sim(r\,\overline{\supset}\,q)\vee\sim p\,\&\sim r$

$=_T \ \sim\,\sim p\,\&\sim q\,\&\sim r\vee\sim p\vee\sim(r\,\&\sim q)\vee\sim p\,\&\sim r$

$=_T \ p\,\&\sim q\,\&\sim r\vee\sim p\vee\sim r\vee\sim\,\sim q\vee\sim p\,\&\sim r$

$=_T \ p\,\&\sim q\,\&\sim r\vee\sim p\vee\sim r\vee q\vee\sim p\,\&\sim r$

$=_T \ (p\vee\sim p\vee\sim r\vee q\vee\sim p)\,\&\,(p\vee\sim p\vee\sim r\vee q\vee\sim r)$
$\quad\&\,(\sim q\vee\sim p\vee\sim r\vee q\vee\sim p)\,\&\,(\sim q\vee\sim p\vee\sim r\vee q\vee\sim r)$
$\quad\&\,(\sim r\vee\sim p\vee\sim r\vee q\vee\sim p)\,\&\,(\sim r\vee\sim p\vee\sim r\vee q\vee\sim r)$

In this particular case the reduction to disjunctive normal form could be carried out simply by omitting the last step, since no uses of (4B) or (5B) are required.

Reductions to normal forms are by no means unique and, in the above example, we can see that the formula $\sim p \vee \sim r \vee q$ would provide a simpler disjunctive (and therefore conjunctive) normal form since the disjuncts $p \& \sim q \& \sim r$, $\sim p \& \sim r$ take the truth value F whenever the disjunct $\sim r$ takes the truth value F. A simpler normal form will, in general, give rise to a simpler electrical circuit so that the problem of finding the simplest normal form of a formula is of practical importance. This problem will be discussed further in Chapter 5.

Examples 2G

1. Reduce the formula
$$p \& q \vee r \supset p \& (q \vee r) \vee [\![q/(p \downarrow s)]\!]$$
   to (i) disjunctive normal form, (ii) conjunctive normal form.
2. Simplify directly the conjunctive normal form obtained at the end of Section 2.7 to $\sim p \vee \sim r \vee q$.
3. Simplify as much as is obviously possible, the results of question 1.
4. Reduce the formulae (a) $p \vee q \equiv [\![p/(q \not\supset r)]\!]$ (b) $(p \supset q) \supset [\![(q/r) \supset (p \not\equiv r \vee s)]\!]$ to (i) disjunctive normal form, (ii) conjunctive normal form.

Solutions 2G

1. (i) $p \& q \vee r \supset p \& (q \vee r) \vee [\![q/(p \downarrow s)]\!]$

$=_T \sim (p \& q \vee r) \vee p \& (q \vee r) \vee \sim q \vee \sim (\sim p \& \sim s)$

$=_T \sim (p \& q) \& \sim r \vee p \& (q \vee r) \vee \sim q \vee \sim \sim p \vee \sim \sim s$

$=_T (\sim p \vee \sim q) \& \sim r \vee p \& (q \vee r) \vee \sim q \vee p \vee s$

$=_T \sim p \& \sim r \vee \sim q \& \sim r \vee p \& q \vee p \& r \vee \sim q \vee p \vee s$

2. Since $p \vee \sim p$, $\sim q \vee q$ are tautologies, the first four disjuncts are tautologies and may be omitted. Hence, by idempotence,

$(\sim r \vee \sim p \vee \sim r \vee q \vee \sim p) \& (\sim r \vee \sim p \vee \sim r \vee q \vee \sim r)$

$=_T (\sim r \vee \sim p \vee q) \& (\sim r \vee \sim p \vee q)$

$=_T \sim r \vee \sim p \vee q$

4. (a) (i) $p \vee q \equiv [\![p/(q \not\supset r)]\!]$

$=_T (p \vee q) \& [\![p/(q \not\supset r)]\!] \vee \sim (p \vee q) \& \sim [\![p/(q \not\supset r)]\!]$

$=_T (p \vee q) \& (\sim p \vee \sim q \vee r) \vee \sim p \& \sim q \& p \& q \& \sim r$

$=_T p \& \sim p \vee p \& \sim q \vee p \& r \vee q \& \sim p \vee q \& \sim q \vee r \vee \sim p \& \sim q \& p \& q \& \sim r$

$=_T p \& \sim q \vee p \& r \vee q \& \sim p \vee q \& r$ (deleting absurdities)

Chapter 3

Logical Computers

3.1 The machine of McCallum and Smith[2, 13]

We have seen in Chapter 1 how, given any formula P of the propositional calculus, we can combine decision elements for the functors occurring in P to construct a decision mechanism for that formula. This construction enables us to use a machine to solve either of the following problems (except, of course, in the case where P is an absurdity).

I. Find an assignment of truth values to the propositional variables of P under which P takes the truth value T.

II. Find all assignments of truth values to the propositional variables of P under which P takes the truth value T.

As certain practical problems are reducible to these two forms, such a machine was constructed by McCallum and Smith as part of an attempt to solve the traffic problem at London Airport. This machine contained 32 decision elements for particular functors and a changing switch which could be used so that, if some or all of these 32 elements were used to form a decision mechanism for a formula containing n ($n \leqslant 7$) propositional variables, the inputs corresponding to these variables could assume, successively, all the 2^n possible n-tuples of physical states. When the output of the decision mechanism corresponded to the truth value T the switch governing the changes of input states was automatically stopped and the n-tuple of states could be recognized by observing the colours of lights next to the respective inputs. If problem I was to be considered the work was then complete, but if the problem was II, the machine was then restarted and the next solution was recorded, the process being continued until all the 2^n cases had been tested.

For example, if the formula P whose decision mechanism was constructed was

$$p \vee q \not\equiv r \mathbin{\&} {\sim} p \quad (n = 3)$$

the output would correspond to the truth value T, causing the machine to stop, in those cases where the respective truth values of the propositional variables p, q, r were

$$T, T, T \qquad T, F, F$$

$$T, T, F \qquad F, T, F$$

$$T, F, T \qquad F, F, T$$

and each of these triples could be recorded by noting the colours of the lights corresponding to p, q, r. Since $2^n = 8$ the machine would test the outputs corresponding to 8 triples of inputs and these outputs would correspond to the truth value T in 6 of the 8 cases.

3.2 The Nottingham University logical computer

The machine of McCallum and Smith would not, of course, cope with a formula containing more than 7 distinct propositional variables or with a formula containing more occurrences of a particular functor than the number of decision elements for that functor. A somewhat more elaborate version of the machine was constructed at Nottingham by Dr. E. Foxley. This recorded solutions and restarted the machine automatically and contained, in addition to a few decision elements for functors $C_i(, ...,)$, $D_i(, ...,)$, 84 decision elements of a special kind. These were known as 'universal decision elements', since they could be used as decision elements for any of the 8 non-trivial binary functors, simply by using 8 positions of an 11-position switch.

These universal decision elements[5, 14, 15] were basically decision elements for the sixth-degree functor $\Phi(, , , , ,)$ whose truth table is determined by the equation

$$\Phi(P, Q, R, S, U, V) =_{\mathrm{T}} [P, (Q \vee R) \& (S/U), V]$$

In the propositional calculus whose primitives are the functor $\Phi(, , , , ,)$ and the logical constants t, f we may (reasonably) define all the non-trivial binary functors. Thus, given a supply of decision elements for $\Phi(, , , , ,)$, we may simulate all non-trivial binary decision elements, the number of decision elements for $\Phi(, , , , ,)$ required being the number of occurrences of $\Phi(, , , , ,)$ on the right-hand side of the definition in question. Since we may, in fact, make appropriate definitions of these 8 functors, using the functor $\Phi(, , , , ,)$ only once in each of the 8 cases, a single decision element for $\Phi(, , , , ,)$ will simulate decision elements for the 8 functors. We may, for example, make the following 8 definitions.

$$P \vee Q =_{df} \Phi(t, P, Q, f, f, f)$$

$$P \& Q =_{df} \Phi(f, f, t, P, Q, t)$$

$$P \supset Q =_{df} \Phi(f, f, P, Q, t, t)$$

$$P \not\supset Q =_{df} \Phi(t, f, P, Q, t, f)$$

$$P \equiv Q =_{df} \Phi(f, P, Q, P, Q, t)$$

$$P \not\equiv Q =_{df} \Phi(t, P, Q, P, Q, f)$$

$$P/Q =_{df} \Phi(t, f, t, P, Q, f)$$

$$P \downarrow Q =_{df} \Phi(f, P, Q, f, f, t)$$

If, for example, a decision element for disjunction is to be simulated, the second and third inputs are used as the two inputs of the simulated element and the remaining inputs are set in fixed ways. All these connexions are made by setting the 11-position switch in the appropriate position. The remaining 3 positions of the switch relate to the simulation of decision elements for 3 ternary functors, one of which is closely connected with a device known as a flip-flop (cf. Section 4.5).

Examples 3A

1. A logical computer contains 19 decision elements (not universal) whose functions are divided as follows (the notation being that of Łukasiewicz).

$$A3 \quad K3 \quad C4 \quad E2 \quad E'3 \quad N4$$

and it can supply inputs corresponding to n ($n \leqslant 5$) propositional variables. Which of the following formulae can be tested by the machine?
   (i) *AKCpqErsCEuqNr*   (ii) *AAACKpqrsNuCvNKpq*
   (iii) *AKKCpqNrNKpsCEqsKpNr*   (iv) *CCpqCKrApsNE'pr*
   (v) *E'NKpNqCKpNrNCps*
2. Prove that the universal decision element of Section 3.2 can simulate decision elements for the functors $C_3(, ,)$, $D_3(, ,)$.
3. Prove that if

$$\Phi(P, Q, R, S, U) =_T [P, (Q \not\equiv R) \& (R \not\equiv S), U]$$

then the functor $\Phi(, , , ,)$ corresponds to a universal decision element and design a corresponding relay circuit.

Solutions 3A

1. (i) Propositional variables occurring are p, q, r, s, u. Hence $n \leqslant 5$ (since $n = 5$). Numbers of occurrences of functors are as follows.

$$A, 1 \leqslant 3 \quad C, 2 \leqslant 4 \quad E, 2 \leqslant 2 \quad K, 1 \leqslant 3 \quad N, 1 \leqslant 4$$

Hence the formula may be tested.

(ii) Propositional variables occurring are p, q, r, s, u, v. Hence $n = 6 > 5$ and the formula cannot be tested.

2. $C_3(P, Q, R) =_{\text{df}} \Phi(f, f, t, P, Q, R)$

3. $P \lor Q =_{\text{df}} \Phi(f, P, t, Q, t)$

                     $P \,\&\, Q =_{\text{df}} \Phi(t, P, f, Q, f)$

Solution 3A, 3

The relay windings must be such that currents flowing from R to Q and S cause the same type of magnetism, both currents being necessary to operate the switch. There must be no residual magnetism when the currents cease to flow.

3.3 Simple problems

Before considering examples of the uses of a logical computer it will be convenient to extend the results of questions 3 and 4 of Examples 2A. The theorem to be proved is known as the Łesniewski–Mihailescu theorem[16, 17] and it can be proved by an extension of the method used in Solutions 2A. We shall give here an alternative proof, leaving the extension of the old method to the reader, as an exercise.

THEOREM. *A formula of the E–N propositional calculus is a tautology if and only if every propositional variable which occurs in it occurs an even number of times and the negation functor also occurs in it, if at all, an even number of times.*

Let $\Phi(P_1, ..., P_n)$ be a formula of this propositional calculus containing no occurrences of any propositional variables other than those denoted by the syntactical variables $P_1, P_2, ..., P_n$. Let us suppose further that $N, P_1, ..., P_n$ occur in $\Phi(P_1, ..., P_n)$ $a, b_1, ..., b_n$ times respectively ($a, b_1, ..., b_n \geqslant 0$) and let us rename the truth values T, F as 0, 1 respectively. (When numerical truth values are used the opposite convention is more common, but it causes complications here.)

LEMMA. *If* $P_1, ..., P_n$, $\Phi(P_1, ..., P_n)$ *take the truth values* $x_1, ..., x_n$, $\phi(x_1, ..., x_n)$ *respectively then*

$$\phi(x_1, ..., x_n) \equiv a + \sum_{i=1}^{n} b_i x_i \quad (\text{mod } 2)$$

We shall prove the lemma by strong induction on the length, l, of $\Phi(P_1, ..., P_n)$. If $l = 1$ then $\Phi(P_1, ..., P_n)$ is, for some integer j, P_j. Thus $a, b_1, ..., b_{j-1}, b_{j+1}, ..., b_n = 0$, $b_j = 1$, $\phi(x_1, ..., x_n) = x_j$ and the lemma is trivial.

We now assume the lemma for $1, ..., l-1$ and deduce it for l. Since we have disposed of the case $l = 1$ we may assume, in future, that $l \geqslant 2$ and it follows at once that $\Phi(P_1, ..., P_n)$ is of one of the forms

$$N\Psi(P_1, ..., P_n) \quad E\Psi(P_1, ..., P_n) \Lambda(P_1, ..., P_n)$$

where no propositional variables other than those denoted by $P_1, ..., P_n$ occur in either of the formulae $\Psi(P_1, ..., P_n)$, $\Lambda(P_1, ..., P_n)$. Let us suppose that, if $P_1, ..., P_n$ take the truth values $x_1, ..., x_n$ respectively, $\Psi(P_1, ..., P_n)$, $\Lambda(P_1, ..., P_n)$ take the truth values $\psi(x_1, ..., x_n)$, $\lambda(x_1, ..., x_n)$ respectively and that N, P_i occur a', b_i' times respectively in $\Psi(P_1, ..., P_n)$ and a'', b_i'' times respectively in $\Lambda(P_1, ..., P_n)$ $(i = 1, ..., n)$.

In the first case

$$l(\Phi) = 1 + l(\Psi)$$

so that

$$l(\Psi) < l(\Phi)$$

and the induction hypothesis applies to $\Psi(P_1, ..., P_n)$. Hence

$$\psi(x_1, ..., x_n) \equiv a' + \sum_{i=1}^{n} b_i' x_i \quad (\text{mod } 2) \tag{1}$$

Rewriting the negation truth table with $0, 1$ in place of T, F respectively we see that

P	NP
0	1
1	0

$\phi(x_1, ..., x_n) \equiv 1 + \psi(x_1, ..., x_n) \quad (\text{mod } 2)$

so that, by (1),

$$\phi(x_1, ..., x_n) \equiv a' + 1 + \sum_{i=1}^{n} b_i' x_i \quad (\text{mod } 2)$$

Since

$$a = a' + 1$$

$$b_i = b_i' \quad (i = 1, ..., n)$$

it follows at once that

$$\phi(x_1, \ldots, x_n) \equiv a + \sum_{i=1}^{n} b_i x_i \quad (\text{mod } 2)$$

In the second case

$$l(\Phi) = l(\Psi') + l(\Lambda) + 1$$

$$\phi(x_1, \ldots, x_n) \equiv \psi(x_1, \ldots, x_n) + \lambda(x_1, \ldots, x_n) \quad (\text{mod } 2)$$

EPQ	0	1	Q
0	0	1	
1	1	0	
P			

$$a = a' + a''$$
$$b_i = b_i' + b_i'' \quad (i = 1, \ldots, n)$$

It then follows, by arguments similar to those used in the first case that

$$\phi(x_1, \ldots, x_n) \equiv a + \sum_{i=1}^{n} b_i x_i \quad (\text{mod } 2)$$

(The details of this part of the argument are left to the reader, as an exercise.) This completes the induction step and the lemma is therefore proved.

Proof of the main theorem. Let the propositional variables of the formula in question be denoted by $P_1, \ldots, P_n$ and let the number of occurrences of $N, P_1, \ldots, P_n$ in the formula be $a, b_1, \ldots, b_n$ respectively $(a \geqslant 0; b_1, \ldots, b_n > 0)$.

If $a, b_1, \ldots, b_n$ are all even then, with the notation of the lemma,

$$\phi(x_1, \ldots, x_n) \equiv 0 \quad (\text{mod } 2)$$

from which it follows (since $\phi(x_1, \ldots, x_n) \in (0, 1)$) that $\phi(x_1, \ldots, x_n) = 0$, irrespective of the values of $x_1, \ldots, x_n$. Thus $\Phi(P_1, \ldots, P_n)$ is a tautology.

If, conversely, $\Phi(P_1, \ldots, P_n)$ is a tautology then

$$0 = \phi(0, \ldots, 0) \equiv a \quad (\text{mod } 2)$$

Thus a is even and

$$0 = \phi(x_1, \ldots, x_n) \equiv \sum_{i=1}^{n} b_i x_i \quad (\text{mod } 2), \quad \text{for all } x_1, \ldots, x_n.$$

In particular, if $x_1, \ldots, x_{j-1}, x_{j+1}, \ldots, x_n = 0$ and $x_j = 1$ then

$$0 \equiv b_j \quad (\text{mod } 2)$$

Thus b_j is even $(j = 1, \ldots, n)$ and the theorem is proved.

Having proved the Łesniewski–Mihailescu theorem let us consider first the very simple problem of the University of Blanktown whose admission requirements are only two.

I. All candidates must have a G.C.E. pass in English or in Mathematics.

II. Candidates wishing to read Agriculture must have a G.C.E. pass in Chemistry.

Determine which candidates can be admitted, given G.C.E. results in English, Mathematics and Chemistry and the subject to be read.

Let us assign values as follows to the propositional variables p, q, r, s.

p: the candidate has a G.C.E. pass in English.

q: the candidate has a G.C.E. pass in Mathematics.

r: the candidate has a G.C.E. pass in Chemistry.

s: the candidate wishes to read Agriculture.

Clearly Condition I is satisfied if and only if at least one of the propositional variables p, q takes the truth value T, i.e. if and only if the formula $p \vee q$ takes the truth value T. Similarly condition II is satisfied if and only if the formula $s \supset r$ takes the truth value T. Thus a candidate is admissible if and only if both the formulae

$$p \vee q \quad s \supset r$$

take the truth value T, i.e. if and only if the formula

$$(p \vee q) \& (s \supset r)$$

takes the truth value T. The machine must therefore test this latter formula.

As a somewhat more complex example let us consider the following problem.

It is known that mathematicians always tell the truth to other mathematicians and lies to physicists and that physicists always tell the truth to other physicists and lies to mathematicians. If

   (i) A tells B that C is a physicist,

   (ii) A tells D that B is a mathematician,

   (iii) B tells E that A and C have different occupations,

   (iv) C tells D that A is a physicist,

   (v) D tells C that B is a mathematician,

and all of A, B, C, D, E are either mathematicians or physicists, how many of them are mathematicians?

Let us assign values to the propositional variables p, q, r, s, u as follows.

p: A is a mathematician.

q: B is a mathematician.

r: C is a mathematician.

s: D is a mathematician.

u: E is a mathematician.

We note that all the statements (i)–(v) are of the form

$$\text{`}\alpha \text{ tells } \beta \text{ that } P\text{'}$$

where

$$\alpha, \beta \in \{A, B, C, D, E\}$$

Let the syntactical variable $Q(R)$ denote the propositional variable whose value is the statement '$\alpha(\beta)$ is a mathematician'. The statement 'α tells β that P' will take the same truth value as P if and only if α and β have the same occupation, i.e. if and only if Q and R take the same truth value or $Q \equiv R$ takes the truth value T. Thus 'α tells β that P' takes the same truth value as the formula $(Q \equiv R) \equiv P$. Since the statement 'α is a mathematician (physicist)' will take the truth value T if and only if the formula $Q (\sim Q)$ takes the truth value T, the statements (i)–(v) may be represented by the respective formulae

$$(p \equiv q) \equiv \sim r \quad (p \equiv s) \equiv q \quad (q \equiv u) \equiv (p \equiv \sim r)$$

$$(r \equiv s) \equiv \sim p \quad (s \equiv r) \equiv q$$

Thus we must determine how many of the propositional variables p, q, r, s, u take the truth value T when the formula

$$C_5(\{p \equiv q\} \equiv \sim r, \quad \{p \equiv s\} \equiv q, \quad \{q \equiv u\} \equiv \{p \equiv \sim r\}, \quad \{r \equiv s\} \equiv \sim p, \quad \{s \equiv r\} \equiv q)$$

takes the truth value T. There is, of course, no *a priori* reason for assuming that the number of propositional variables which take the truth value T in this case is unique but, in fact, whenever the given formula takes the truth value T, exactly 3 of these propositional variables take the truth value T.

We may use the lemma of the Łesniewski–Mihailescu theorem to check that, on all occasions where the output of the corresponding decision mechanism corresponds to T, exactly 3 of the inputs relating to p, q, r, s, u will correspond to the truth value T. If the given formula takes the truth value T then

$$(p \equiv q) \equiv \sim r \text{ takes the truth value } T \tag{1}$$

$$(p \equiv s) \equiv q \text{ takes the truth value } T \tag{2}$$

$$(q \equiv u) \equiv (p \equiv \sim r) \text{ takes the truth value } T \tag{3}$$

$$(r \equiv s) \equiv \sim p \text{ takes the truth value } T \tag{4}$$

$$(s \equiv r) \equiv q \text{ takes the truth value } T \tag{5}$$

By (1), (2) and the lemma, the formulae

$$[\![(p \equiv q) \equiv \sim r]\!] \equiv [\![(p \equiv s) \equiv q]\!] \quad \sim r \equiv s$$

4

take the truth value T (since if $p, q, \sim r, s$ take the truth values x, y, z, w respectively then

$$2x + 2y + z + w \equiv z + w \quad (\text{mod } 2)$$

so that

$$r \neq_T s \tag{6}$$

By (4) and (6)

$$p \text{ takes the truth value } T \tag{7}$$

By (1), (7) and the lemma

$$q \neq_T r \tag{8}$$

By (3), (7), (8) and the lemma

$$u \text{ takes the truth value } T \tag{9}$$

By (5) and (6)

$$q \text{ takes the truth value } F \tag{10}$$

By (8) and (10)

$$r \text{ takes the truth value } T \tag{11}$$

By (6) and (11)

$$s \text{ takes the truth value } F \tag{12}$$

Thus, by (7), (10), (11), (12) and (9) the only possible solution is that where p, q, r, s, u take the truth values T, F, T, F, T respectively and it is easily checked that, in this case, the tested formula takes the truth value T.

As a further example let us consider the problem of determining the truth tables of those ternary functors $\Phi(\,,\,,\,)$ such that disjunction is (reasonably) definable in terms of $\Phi(\,,\,,\,)$, t and f in such a way that the functor $\Phi(\,,\,,\,)$ occurs on the right-hand side of the definition only once.

If

$$U, V, W \in (P, Q, t, f)$$

and

$$P' = Q \quad Q' = P \quad t' = t \quad f' = f$$

and we may properly make the definition

$$P \vee Q =_{\mathrm{df}} \Phi(U, V, W)$$

then, by the commutativity of the disjunction truth table, an alternative (and reasonable) definition would be

$$P \vee Q =_{\mathrm{df}} \Phi(U', V', W')$$

Hence, if disjunction may be properly defined, this may be done in such a way that the first occurrence of P on the right-hand side of the definition precedes the first such occurrence of Q when we define $P \vee Q$. Thus disjunction is reasonably definable if and only if at least one of the following 9 formulae has the same truth table as $P \vee Q$.

$$\Phi(P,Q,t) \quad \Phi(P,Q,f) \quad \Phi(P,t,Q)$$

$$\Phi(P,f,Q) \quad \Phi(t,P,Q) \quad \Phi(f,P,Q)$$

$$\Phi(P,P,Q) \quad \Phi(P,Q,P) \quad \Phi(P,Q,Q)$$

Let R_i be a formula which takes the truth value T if and only if the ith of these 9 formulae has the same truth table as $P \vee Q$. The formula to be tested is then

$$D_9(R_1, ..., R_9)$$

We shall show how to construct the formula R_1, the methods for the formulae $R_2, ..., R_9$ being very similar.

Let $\Phi(X,Y,Z)$ take the truth value y_k when X, Y, Z take the truth values x_1, x_2, x_3 respectively (T, F being renamed 1, 0 respectively) and

$$k = x_1 + 2x_2 + 4x_3 + 1$$

and let the value of the propositional variable p_k be the statement $y_k = 1$ (or $y_k = T$) ($k = 1, ..., 8$). The formula $\Phi(P,Q,t)$ will have the same truth table as $P \vee Q$ if and only if the formulae $\Phi(t,t,t)$, $\Phi(f,t,t)$, $\Phi(t,f,t)$, $\Phi(f,f,t)$ take the truth values T, T, T, F respectively, i.e. if and only if p_8, p_7, p_6, p_5 take the truth values T, T, T, F respectively. Thus we may take for R_1 the formula

$$C_3(p_6, p_7, p_8) \mathbin{\not\!\supset} p_5$$

Examples 3B

1. It is known that salesmen always tell the truth and engineers are all liars. A, ..., G are all salesmen or engineers.

   A says that B states that C ridicules that D declares that E denies that F affirms that G is a salesman.

   B and D are salesmen and F is an engineer.

   G says E is an engineer.

   A and C are not both salesmen.

   Explain how to use a logical computer to determine the number of salesmen and give also a purely mathematical solution to the problem.

2. Determine, for the last problem of Section 3.3, the formulae R_2 and R_9.

3. Explain how to solve the problems corresponding to that of the last part of Section 3.3 for the conjunction and material implication functors.

Solutions 3B

1. Let the value of p be the statement 'A is a salesman' and let q, r, s, u, v, w relate similarly to B, C, D, E, F, G respectively. Thus the tested formula is the conjunction of the formulae

$$p \equiv \{q \equiv \| r \equiv \sim \langle s \equiv [\![u \equiv \sim (v \equiv w)]\!]\rangle \|\} \quad q, s, \sim v$$

$$w \not\equiv u \quad p/r$$

Since $q, s, \sim v$ take the truth value T it follows from the formula corresponding to the first condition's taking the truth value T, and the lemma of the Łesniewski–Mihailescu theorem, that the formula

$$p \equiv [\![r \equiv (u \equiv w)]\!]$$

takes the truth value F and, since $w \not\equiv u$ takes the truth value T, $p \equiv r$ takes the truth value T. Since p/r also takes the truth value T, p, r both take the truth value F. Thus p, q, r, s, v take the truth values F, T, F, T, F respectively and u, w take opposite truth values. Hence exactly 3 of A, ..., G are salesmen. (It can easily be checked that both solutions satisfy the conditions, though one solution checked is sufficient to establish that there are exactly 3 salesmen.)

2. R_2 may be taken as $C_3(p_2, p_3, p_4) \not\supset p_1$ and R_9 as $\sim p_1 \& p_2 \& p_7 \& p_8$ or $C_3(p_2, p_7, p_8) \not\supset p_1$.

3. For material implication, a definition is possible if and only if a 'leading P' definition is possible for at least one of the functors $\supset, \subset$. Thus a suitable formula corresponding to R_1 is

$$p_5 \& p_6 \& \sim p_7 \& p_8 \lor p_5 \& \sim p_6 \& p_7 \& p_8$$

or

$$C_3(p_5, p_8, p_6 \not\equiv p_7)$$

The 8 remaining disjuncts may be found similarly.

Chapter 4

The Construction of Certain Types of Computing Machinery

4.1 Digital computer components and their use in combination

The logical computers discussed in Chapter 3 consisted mainly of a single mechanism, namely, a decision mechanism for a given formula of the propositional calculus, which was used repeatedly to determine the truth value of that formula under various assignments of truth values to the propositional variables occurring in it. In a modern digital computer we are, however, concerned with the carrying out of a complex calculation in which, in order to obtain the final result or results, many evaluations of different types have to be performed in succession. In general the calculations are carried out one at a time, though in rapid succession, but some computers are capable of running through several sequences of calculations simultaneously.

As a simple example, let us consider the evaluation of t, where

$$t = (a+b)(c^d+e!)$$

given the values (as non-negative integers) of a, b, c, d, e. The machine would receive a number of instructions as follows.

Instruction number	Operation	First argument	Second argument (if any)	Result
1	Add	a	b	$x(x = a+b$ by entries in the 3 previous columns)
2	Exponentiate	c	d	y
3	Form factorial	e	—	z
4	Add	y	z	w
5	Multiply	x	w	t

89

Although the instructions $1, 2, 3, 4, 5$ could be obeyed successively, it is not always necessary for a particular stage to be completed before the next one is begun, as may be seen from Figure 4.1. (The actual numbers used in

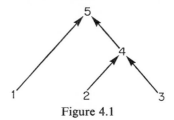

Figure 4.1

Figure 4.1 relate, of course, to instructions from the above table.) Instructions 1, 2, 3 may all be obeyed immediately. As soon as instructions 2, 3 have been obeyed the machine may carry out instruction 4 after which (as instruction 1 will now have been obeyed) the machine may obey instruction 5 and obtain the required answer.

The machine has fed into it inputs corresponding to the values of a, b, c, d, e and gives an output corresponding to the value of

$$t \qquad \text{where } t = (a+b)(c^d+e!)$$

These inputs are sometimes punched on cards and read into the computer, though other methods of entry are also used. Outputs (one only in the above simple example) may be supplied in a similar manner, and many machines can deal with several inputs and outputs simultaneously.

The numbers initially supplied to the machine (a, b, c, d, e in the above example) and the numbers evaluated as intermediate results (x, y, z, w) are fed into a special mechanism, known as the *memory unit*, as soon as they are obtained. We note that, in our example, the last four instructions all relate to different operations, and certain parts of the machine which deal with these and other similar operations together form what is known as the *arithmetic unit*. A third unit, known as the *control unit*, governs the sequential execution of the machine's instructions, by interpreting the instructions and telling the arithmetic unit what to do next. The various parts of the memory unit, which can store one number each, themselves have numbers assigned to them. These latter numbers are known as *addresses*. Instructions are normally worded with respect to addresses, so that, for instance, in our example the fourth instruction could begin 'Add the contents of address 17 to those of address 24'. Instructions are themselves in the form of binary numbers, located at addresses, so the above instruction might be completed as follows: 'Write the result of the addition at address 31 and then obey the instruction at address 5'. (We are here supposing that instructions $1, ..., 5$ are located at addresses $1, ..., 5$ respectively.) The ideas of this section are discussed much more fully by other authors, such as Ledley.[9] The computer may be instructed to carry out calculations which amount to the simulation of a logical computer (cf. Chapter 3). The details of possible methods are often given in computer programming manuals.

The design of the units described here makes considerable use of the theory of logical decision mechanisms. The remainder of this chapter will be devoted to discussions of the construction of certain such mechanisms and further examples will be considered later in the book.

4.2 Relevance and limitation of general approaches to the analysis of logical circuitry

In a digital computer it is usual to use the binary number system, so that, corresponding to the everyday method of writing the number equal to

$$\sum_{i=1}^{n} 10^{i-1} x_i [\![x_1, ..., x_n \in \{0, 1, ..., 9\}, x_n \neq 0]\!]$$

as

$$x_n x_{n-1} \cdots x_1$$

the number equal to

$$\sum_{i=1}^{n} 2^{i-1} x_i [\![x_1, ..., x_n \in \{0, 1\}, x_n = 1]\!]$$

is written as

$$x_n x_{n-1} \cdots x_1$$

Thus, for example, the decimal number 97 is equal to

$$64 + 32 + 1$$

or

$$(2^6 . 1) + (2^5 . 1) + (2^4 . 0) + (2^3 . 0) + (2^2 . 0) + (2^1 . 0) + (2^0 . 1)$$

so that the corresponding binary representation of this number is

$$1100001$$

The two distinct digits of the binary number system may then be represented by two distinct physical states of a wire and these states (cf. Chapter 1) may, in turn, be regarded as representing the truth values T, F. Thus, as we shall show in more detail in this and subsequent chapters, problems concerning the construction of computing machinery may, to a considerable extent, be reduced to problems in mathematical logic. We shall see, in this and later chapters, that extra-logical considerations are necessary in view of the need to allow for time delays. It is, of course, essential also to take into account the different methods of physical representation of different logical functors and their relative costs. It is also vital that the logic problem shall not be impossibly complicated.

4.3 The serial adder

Let us now consider the construction of a particular type of mechanism for adding two non-negative integers. We shall, as we explained in Section 4.2, use the binary number system. Let us denote the two summands by x, y, where

$$x = \sum_{i=1}^{m_1} 2^{i-1} x_i \quad y = \sum_{i=1}^{m_2} 2^{i-1} y_i \quad x_{m_1} = y_{m_2} = 1$$

and

$$x_1, \ldots, x_{m_1-1} \quad y_1, \ldots, y_{m_2-1} \in \{0, 1\}$$

It will be convenient to make the definition

$$n = \max(m_1, m_2)$$

and write

$$x = \sum_{i=1}^{n} 2^{i-1} x_i \quad y = \sum_{i=1}^{n} 2^{i-1} y_i$$

where

$$x_{m_1+1}, \ldots, x_n \quad y_{m_2+1}, \ldots, y_n = 0$$

so that the 'leading 1' convention of Section 4.2 is discarded, 'leading zeros' now being permitted. (If $\alpha > \beta$ the sequence $w_\alpha, \ldots, w_\beta$ is regarded as vacuous.)

If $x + y = z$ then the binary representation of z cannot contain more than $n + 1$ digits, so we may write

$$z = \sum_{i=1}^{n+1} 2^{i-1} z_i$$

where

$$z_1, \ldots, z_{n+1} \in \{0, 1\}$$

(It follows easily that if $z_{n+1} = 0$, i.e. if z has a leading zero, then $z_n = 1$.)

We shall show how to construct a mechanism, known as a 'serial adder' with two input wires and one output wire. The first input wire is supplied, successively, with inputs corresponding to the values of the digits $x_1, x_2, \ldots, x_n$, the length of time elapsing between two successive input signals being constant (cf. Section 1.9). This length of time will be referred to, in future, as a *unit.* The second input wire is supplied with inputs corresponding to the values of $y_1, y_2, ..., y_n$ at intervals of one unit, the inputs corresponding to x_1, y_1 being supplied to the two wires simultaneously. We wish to construct the adder so that the output wire supplies, in succession, signals corresponding to the values of $z_1, ..., z_{n+1}$. Let us suppose that the digits 1, 0 have the same physical representations as the truth values T, F respectively.

As regards manual addition, two binary numbers may be added by the conventional 'sum and carry' method commonly used for decimal numbers. For example, the evaluation of

$$97 + 243$$

in decimal and binary notations is shown below, digits to be carried from a column to the next being written between the columns and leading zeros being introduced as explained above.

```
    0 1 1           1 1 1 0 0 0 1 1
    0 9 7           0 1 1 0 0 0 0 1
    2 4 3           1 1 1 1 0 0 1 1
    ─────           ───────────────
    0 3 4 0         1 0 1 0 1 0 1 0 0
```

(We note that $243 = 128 + 64 + 32 + 16 + 2 + 1$, $340 = 256 + 64 + 16 + 4$.)

Let us denote by c_i the digit carried after dealing with the columns containing x_i and y_i $(i = 1, ..., n)$. When we deal with this column we add the three numbers

$$x_i \quad y_i \quad c_{i-1}$$

$(i = 1, ..., n; c_0 = 0)$, each of these numbers being 0 or 1. Thus, for all i $(i = 1, ..., n)$, if

$$w_i = x_i + y_i + c_{i-1} \tag{1}$$

then

$$w_i \in \{0, 1, 2, 3\}$$

or, in a binary notation (with a possible leading zero),

$$w_i \in \{00, 01, 10, 11\}$$

Clearly the value of w_i determines the values of z_i and c_i uniquely. $z_i = 1$ if and only if $w_i \in \{01, 11\}$ and $c_i = 1$ if and only if $w_i \in \{10, 11\}$. Since the value of w_i is 0, 1, 2 or 3 according as 0, 1, 2 or 3 of x_i, y_i, c_{i-1} are equal to 1 (cf. (1)) it follows at once that

$z_i = 1$ if and only if exactly 1 or 3 of x_i, y_i, c_{i-1} is equal to 1,   (A)

$c_i = 1$ if and only if at least 2 of x_i, y_i, c_{i-1} are equal to 1,   (B)

$(i = 1, \ldots, n)$. Since we may define $x_{n+1} = y_{n+1} = 0$, thereby introducing (possibly additional) leading zeros for x, y, the statements (A), (B) remain valid for $i = n+1$ though, of course, $c_{n+1} = 0$.

Let the values of the digits x_i, y_i, z_i, c_j correspond to the states of wires which correspond, in turn, to the truth values of the propositional variables p_i, q_i and the formulae P_i, Q_j respectively $(i = 1, \ldots, n+1; j = 0, \ldots, n+1)$. We must therefore find suitable formulae

$$P_i, Q_j \quad (i = 1, \ldots, n+1; j = 0, \ldots, n+1)$$

the formulae $Q_0, \ldots, Q_n$ being required to construct $P_1, \ldots, P_{n+1}$ and the formula Q_{n+1} not being strictly required (see below).

By (A), P_i takes the truth value T if and only if exactly 1 or 3 of the formulae p_i, q_i, Q_{i-1} take the truth value T and, by (B), Q_i takes the truth value T if and only if at least 2 of p_i, q_i, Q_{i-1} take the truth value T. Thus, since $c_0 = 0$, we may make the following inductive definitions, where $L_2(\,,\,)$ is a ternary functor whose truth table is such that $L_2(P, Q, R)$ take the truth value T if and only if at least 2 of P, Q, R take the truth value T.

$$Q_0 = f$$
$$P_i = (p_i \equiv q_i) \equiv Q_{i-1} \quad (i = 1, \ldots, n+1)$$
$$Q_i = L_2(p_i, q_i, Q_{i-1}) \quad (i = 1, \ldots, n+1)$$

(The result for P_i follows from the Łesniewski–Mihailescu theorem.)

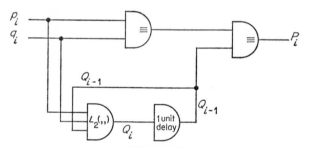

Figure 4.2

A signal representing the formula Q_{i-1} may be obtained from a signal representing Q_i $(i = 1, \ldots, n+1)$ by passing it through a delay mechanism whose output signal is the same as the input except that it occurs one unit of time later. (The numerical value of the suffix will not change, but the value of i increases by 1, so the new value of $i-1$ is the same as the old value of i.) If we use the pulse-and-pause representation of truth values we automatically ensure that Q_0 is correctly represented, so we may use the mechanism illustrated in Figure 4.2. The output corresponding to Q_{n+1} is provided ultimately, but it must, of course, correspond to the truth value F, so it is not really needed.

In practice decision elements for equivalence and $L_2(\ ,\ )$ would not be used, but they would be replaced by equivalent decision elements for other functors. For example, we could take

$$P_i = \llbracket (p_i \& q_i) \vee \sim (p_i \vee q_i) \rrbracket \& Q_{i-1} \vee \sim D_3(p_i \& q_i, \ \sim \llbracket p_i \vee q_i \rrbracket, Q_{i-1})$$

$$Q_i = C_3(p_i \vee q_i, p_i \vee Q_{i-1}, q_i \vee Q_{i-1}) \quad (i = 1, \ldots, n+1)$$

4.4 The parallel adder

An adder which is, in general, much faster than the serial adder is obtained by using $n+1$ adders simultaneously, the output corresponding to Q_i being no longer fed back through a delay unit, but into the adder with an output corresponding to Q_{i+1} $(i = 1, \ldots, n)$. Thus, if $j = i+1$, the jth adder uses inputs corresponding to p_j, q_j, Q_i and its output corresponds to P_j. (For P_1 the Q_0 wire may be left disconnected as $Q_0 =_T f$.) If we consider decision elements to be those for equivalence and $L_2(\ ,\ )$, a parallel adder may then be constructed as shown in Figure 4.3.

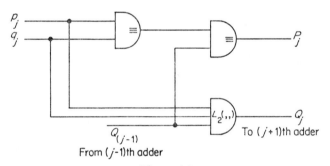

Figure 4.3

Since the outputs of the jth adder depend on the outputs (corresponding to the formulae $Q_1, \ldots, Q_{j-1}$) of the previous adders, each output will be obtained slightly later than any of the previous ones and it is necessary to wait for the whole adder to reach a steady state. The length of time required for this will, of course, depend on the number of times a carry has to be propagated from adders to their successors. As examples let us consider the evaluation, by a parallel adder, of the sums

$$247 + 165 \qquad 453 + 189$$

In binary notation the respective sums become

$$11110111 + 10100101 \qquad 111000101 + 010111101$$

For the first sum the work would proceed as follows, carry digits of 1 being shown immediately above each sum.

```
                                     11110111
                                     10100101
                                     1 1  1 1
First sum                            01010010
                                     1    1
Sum after first propagation          100011000
Sum after second propagation         110011100
```

After two propagations no new carry digits are obtained and the correct sum is 110011100 or, in decimal notation, 412. For the second sum the stages are now as shown.

```
                                     111000101
                                     010111101
                                     1    1 1
First sum                            101111000
                                     1    1
Sum after first propagation          001110010
                                          1
Sum after second propagation         1001100010
                                          1
Sum after third propagation          1001000010
                                          1
Sum after fourth propagation         1000000010
Sum after fifth propagation          1010000010
```

The correct sum of 1010000010 or, in decimal notation, 642 is eventually obtained but, in this case, 5 propagations are necessary.

Examples 4A

1. Using the binary number system evaluate

   (i) $571 + 305$ (ii) $406 + 377$ (iii) $873 + 126$

2. Prove that a serial adder may be manufactured from two decision elements for equivalence, one decision element for conditioned disjunction and a delay unit.
3. Prove that the two equivalence decision elements of question 2 may be replaced by decision elements for non-equivalence.
4. Use the parallel adder method to evaluate the sums of question 1, determining, in each case, how many propagations are required.

Solutions 4A

1. (i)

$$571 = 512 + 32 + 16 + 8 + 2 + 1$$

$$305 = 256 + 32 + 16 + 1$$

```
0 0 0 1 1 0 0 1 1
1 0 0 0 1 1 1 0 1 1
0 1 0 0 1 1 0 0 0 1
```
───────────────────
```
1 1 0 1 1 0 1 1 0 0
```

In decimal notation the sum obtained is equal to

$$512 + 256 + 64 + 32 + 8 + 4$$

or 876.

2. $L_2(P, Q, R) =_T [P, P \equiv Q, R]$

4. (i)

```
1000111011
0100110001
```
──────────
```
                11   1
First sum        1100001010
                          1
Sum after first propagation  1101101000
Sum after second propaga-
tion                          1101101100
```

Two propagations are required.

4.5 High-speed adders

The propagation of carry digits of Section 4.4 may be avoided by the construction of formulae $P_1, P_2, \ldots$ (and the corresponding decision mechanisms) in such a way that the formulae $Q_0, Q_1, \ldots$ are not explicitly involved. Let us, with the notation of Section 4.3, consider the evaluation of

$$z = x + y$$

and let us, to this end, consider the evaluation, for an arbitrary particular value of i $(i \geqslant 1)$ of z_i. Let k $[[= k(i)]]$ be the greatest integer such that

$$k < i \quad x_k = y_k$$

so that

$$x_j \neq y_j \quad (j = k+1, \ldots, i-1)$$

if such an integer exists, and let

$$k = 0$$

if no such integer exists. If $x_k, y_k = 0$ then, with the notation of Section 4.3,

$$c_k = 0$$

and it follows easily that

$$c_{k+1}, \ldots, c_{i-1} = 0$$

Hence

$$z_i = 1 \quad \text{if and only if} \quad x_i \neq y_i$$

If, on the other hand, $x_k, y_k = 1$ then

$$c_k = 1$$

and it follows easily that

$$c_{k+1}, \ldots, c_{i-1} = 1$$

Thus

$$z_i = 1 \quad \text{if and only if} \quad x_i = y_i$$

Hence, in general,

$$z_i = 1$$

if and only if, for some integer k ($0 \leq k < i$, regarding x_0, y_0 as 0), either

$$x_{i-1} \neq y_{i-1}, \ldots, x_{k+1} \neq y_{k+1} \quad x_k = y_k = 0 \quad x_i \neq y_i \tag{A}$$

or

$$x_{i-1} \neq y_{i-1}, \ldots, x_{k+1} \neq y_{k+1} \quad x_k = y_k = 1 \quad x_i = y_i \tag{B}$$

Hence, translating, as in Section 4.3, digital notation to the propositional calculus, we may take (omitting some associations)

$$P_i = \sum_{k=0}^{i-1} \left[\prod_{j=k+1}^{i} (p_j \not\equiv q_j) \,\&\, (p_k \downarrow q_k) \right]$$

$$\vee \sum_{k=0}^{i-2} \left[\prod_{j=k+1}^{i-1} (p_j \not\equiv q_j) \,\&\, (p_i \equiv q_i) \,\&\, p_k \,\&\, q_k \right] \vee (p_i \equiv q_i) \,\&\, p_{i-1} \,\&\, q_{i-1}$$

As an example of addition by the new method let us consider the evaluation of the tenth (in increasing order of significance) digit of the sum

$$101110100001101 + 110011011110101$$

(i.e. $23821 + 26357$ in decimal notation). By the ordinary 'sum-and-carry' method we see that this sum is 1100010000000010 (50178 in decimal notation).

$$
\begin{array}{cccccccccccccccc}
1 & 0 & 1 & 1 & 1 & 0 & 1 & 0 & 0 & 0 & 0 & 1 & 1 & 0 & 1 \\
1 & 1 & 0 & 0 & 1 & 1 & 0 & 1 & 1 & 1 & 1 & 0 & 1 & 0 & 1 \\
\hline
1 & 1 & 1 & 1 & 1 & 1 & 1 & 1 & 1 & 1 & 1 & 1 & 0 & 1 \\
1 & 1 & 0 & 0 & 0 & 1 & 0 & 0 & 0 & 0 & 0 & 0 & 0 & 1 & 0 \\
\end{array}
$$

To evaluate the tenth digit directly (without evaluating $z_1, \ldots, z_9$ first) we note that $x_i \neq y_i$ $(i = 4, \ldots, 9)$, $x_3 = y_3 = 1$. Hence, by (B), since $x_{10} \neq y_{10}$, $z_{10} = 0$. Similarly, since

$$x_{15} = y_{15} \quad x_{14} \neq y_{14} \quad x_{13} \neq y_{13} \quad x_{12} \neq y_{12} \quad x_{11} = y_{11} = 1$$

it follows from (B) that $z_{15} = 1$ and since

$$x_3 = y_3 \quad x_2 = y_2 = 0$$

it follows from (A) that $z_3 = 0$.

Such an adder, although very fast, requires more equipment than the conventional parallel adder, though, as a compromise, fast adders[18] for small maximum values of i may be used in parallel. A similar method of addition is used in the Kilburn[19] adder. Considerable further discussion of this and related problems is given by Ledley.[9]

For example, the above sum could be evaluated by adding digits four at a time and then propagating the carry numbers as follows, the propagated carry digits being shown between the blocks between which they are propagated.

	0101	1101	0000	1101
	0110	0110	1111	0101
	1		1	
First sum	1011	0011	1111	0010
		1		
Sum after first propagation	1100	0011	0000	0010
Sum after second propagation	1100	0100	0000	0010

Examples 4B

1. If

$$x = 1100011010100011010000111010001$$

$$y = 1001100110011010101001000011011$$

evaluate $z_{10}, z_{16}, z_{22}, z_{29}$.

2. Evaluate the sum

$$1111010011011011101 + 1100000111010111001$$

using the method of fast adders for (i) 4 digits, (ii) 5 digits.

Solutions 4B

1. $x_{10} \neq y_{10}, x_9 \neq y_9, x_8 \neq y_8, x_7 \neq y_7, x_6 = y_6 = 0$. Hence, by (A), $z_{10} = 1$.

2. (i)

	0111	1010	0110	1101	1101
	0110	0000	1110	1011	1001

		1	1	1	
First sum	1101	1010	0100	1000	0110
Sum after first propagation	1101	1011	0101	1001	0110

4.6 Subtractors, adder–subtractors and shift registers

Let us suppose that, with the notation of Section 4.3, we now wish to construct a mechanism whose outputs correspond (serially) to the binary number $x - y$, instead of to $x + y$. We shall redefine the number z by the equation

$$z = x - y$$

and suppose, initially, that $x \geqslant y$. The number z may be evaluated manually by an imitation of the normal 'subtractive carry' method for decimal numbers. Thus, in the decimal and binary number systems we may establish that

$$417 - 254 = 163 \quad \text{(i.e. } 110100001 - (0)11111110 = (0)10100011)$$

by the methods shown below.

	4 1 7	1 1 0 1 0 0 0 0 1
	2 5 4	0 1 1 1 1 1 1 1 0
Subtractive carry	1 0	1 1 1 1 1 1 1 0
Difference	1 6 3	0 1 0 1 0 0 0 1 1

Let us denote by c_i the digit carried to the $(i+1)$th column to enable z_i to be evaluated $(i = 0, ..., n)$ where, of course,

$$c_0, c_n = 0$$

Then $z_i = 1$ if and only if

$$x_i - (y_i + c_{i-1}) \equiv 1 \quad (\text{mod } 2)$$

Since all integers α, β satisfy the congruence

$$\alpha + \beta \equiv \alpha - \beta \quad (\text{mod } 2)$$

the above condition is equivalent to the condition

$$x_i + y_i + c_{i-1} \equiv 1 \quad (\text{mod } 2)$$

so that, as in Section 4.3,

$$z_i \equiv x_i + y_i + c_{i-1} \quad (\text{mod } 2)$$

We note that the subtractive carry c_i is equal to 1 if and only if the evaluation of z_i involves new 'borrowing', i.e. if and only if

$$x_i < y_i + c_{i-1}$$

This condition is equivalent to the condition

$$1 - x_i + y_i + c_{i-1} > 1$$

and therefore to the condition

$$1 - x_i + y_i + c_{i-1} \geqslant 2$$

Thus the condition is satisfied if and only if at least two of the integers

$$1 - x_i \quad y_i \quad c_{i-1}$$

are equal to 1.

Hence we may take

$$P_i = (p_i \equiv q_i) \equiv Q_{i-1} \quad (i = 1, \ldots, n) \quad Q_0 = f$$

as for the serial adder and we may also take

$$Q_i = L_2(\sim p_i, q_i, Q_{i-1}) \quad (i = 1, \ldots, n)$$

the sub-formula p_i of the corresponding equation for the serial adder being replaced by $\sim p_i$. It follows at once that the serial adder of Section 4.3 (Figure 4.2) may be converted to a serial subtractor by using an additional decision element for negation as shown in Figure 4.4.

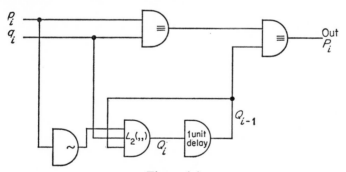

Figure 4.4

If, as a further alternative, we take

$$Q_i = L_2(p_i \equiv r, q_i, Q_{i-1}) \quad (i = 1, \ldots, n+1)$$

then, if r takes the truth value T,

$$Q_i =_T L_2(p_i, q_i, Q_{i-1}) \quad (i = 1, \ldots, n+1)$$

(where Q_{n+1} takes the truth value F but Q_n does not necessarily take this truth value, cf. Section 4.3) and, if r takes the truth value F,

$$Q_i =_T L_2(\sim p_i, q_i, Q_{i-1}) \quad (i = 1, \ldots, n)$$

(the case $i = n+1$ now being irrelevant). Thus we may construct, as shown in Figure 4.5, a combined adder–subtractor by replacing the negation decision

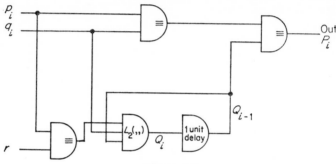

Figure 4.5

element of Figure 4.4 with an equivalence decision element. If the additional input is in a physical state corresponding to the assignment of the truth value T (F) to the propositional variable r then the mechanism acts as a serial adder (subtractor). Similar constructions may, of course, be used for parallel subtractors and adder–subtractors.

We have, up to this point, assumed that $x \geqslant y$. We shall assume, in future, that $x < y$. We note that

$$x - y = -(y - x) \quad y - x > 0$$

and that

$$y - x = 2^n - (2^n + x - y) \quad 2^n + x > y$$

Since $2^n + x > y$ the subtractor will function normally when instructed to subtract y from $2^n + x$ and, since $2^n + x$ differs from x only in the $(n+1)$th digit, the subtractor will supply the first n digits of $2^n + x - y$ when instructed to subtract y from x. Hence, if subsequently instructed to subtract the result from 2^n, it will supply the first n digits of $2^n - (2^n + x - y)$, i.e. of $y - x$. Since 2^n and 0 differ only in the $(n+1)$th digit we may, instead, instruct the subtractor to subtract y from x and then subtract the result from zero, thereby obtaining the first n digits of $y - x$. If z is an n-digit number then the number

$2^n - z$ is known as the *subtractive complement* of z. For example, assigning values to x, y as above, except for interchanging, we obtain 010100011 again.

0 1 1 1 1 1 1 0		0 0 0 0 0 0 0 0 0
1 1 0 1 0 0 0 0 1		1 0 1 0 1 1 1 0 1
1 0 0 0 0 0 0 0 1		1 1 1 1 1 1 1 1 1
1 0 1 0 1 1 1 0 1		0 1 0 1 0 0 0 1 1

If, for some positive integer k, k numbers of not more than n digits (i.e. n-digit numbers which may have leading zeros) are to be added then the partial sums (i.e. the sums of the first l numbers for $l = 1, ..., k-1$) which are first evaluated are stored in a special memory unit known as an *accumulator*. Thus, if the summands are denoted by $\alpha_1, ..., \alpha_k$ and

$$S_l = \sum_{i=1}^{l} \alpha_i \quad (l = 1, ..., k)$$

then S_l is evaluated by adding α_l to S_{l-1} $(l = 2, ..., k)$, S_k being the sum ultimately required. As the inputs of the serial adder determine the digits $x_{l1}, ..., x_{ln}$ of S_l they are transferred to the accumulator and then to one of the input wires of the adder, so that S_{l+1} may be evaluated by supplying the other input wire with digits $y_{l+1,1}, ..., y_{l+1,n}$ of α_{l+1} $(l = 1, ..., k-1)$. (We assign a value to n in such a way that S_k has not more than n digits.)

Each digit x_{lm} $(m = 1, ..., n)$ of S_l, when supplied to the accumulator, occupies the first location of its storage space, each of the digits $x_{l1}, ..., x_{l,m-1}$ being moved one location beyond its immediately previous position. This movement is accomplished by a device known as a *shift register*, consisting of n *flip-flops* or single-digit memory devices. These devices and their use in shift registers will be considered at the end of this chapter.

Examples 4C

1. Find the subtractive complements of
   (i) 1011101010   (ii) 11101100011   (iii) 11011011100010
2. Evaluate $101110101011 - 100111000010$ by the subtractive carry method. Evaluate also the (negative) difference when the roles of the two numbers are interchanged, using the subtractive complement method.
3. Evaluate, by the method of partial sums
   $$101101 + 111010 + 100001 + 1011 + 10100$$
4. Design a serial adder–subtractor using only a delay unit and decision elements for non-equivalence and the functor $L_2(, ,)$.
5. Design a mechanism with inputs corresponding to propositional variables p_i, q_i, r such that the outputs correspond (serially) to the numbers (with the notation of Section 4.6) $x - y, 2x - y$ according as r takes the truth value T or the truth value F, whenever $x \geqslant y$.

Solutions 4C

1. (i)
```
0 0 0 0 0 0 0 0 0 0
1 0 1 1 1 0 1 0 1 0
1 1 1 1 1 1 1 1 0
```
```
0 1 0 0 0 1 0 1 1 0
```

3. (i)
```
    101101
    111010
```
```
   1100111
    100001
```
```
  10001000
      1011
```
```
  10010011
     10100
```
```
  10100111
```

4.
$$P_i = (p_i \equiv q_i) \equiv Q_{i-1} =_T (p_i \not\equiv q_i) \not\equiv Q_{i-1}$$
$$Q_i = L_2(p_i \equiv r, q_i, Q_{i-1}) =_T L_2(\llbracket p_i \not\equiv r \rrbracket \not\equiv t, q_i, Q_{i-1})$$

4.7 The multiplier

Multiplication may be reduced to addition as in the case of manual (decimal or binary) multiplication. For example, the manual evaluations of the product

$$125 \,.\, 103$$

in the two number systems are as follows. (We note that $125 = 64 + 32 + 16 + 8 + 4 + 1$, $103 = 64 + 32 + 4 + 2 + 1$.)

125	1111101
103	1100111
———	———
375	1111101
000	1111101
125	1111101
———	0000000
12875	0000000
	1111101
	1111101
	———
	11001001001011

(We note that $12875 = 8192 + 4096 + 512 + 64 + 8 + 2 + 1$.) In a serial multiplier the method is similar, though the additions are performed individually so that, for example, the product

$$101.110$$

(5.6 in decimal notation) would be evaluated as shown below. (We note that the binary representation of 30 is 11110.)

$$
\begin{array}{r}
101 \\
110 \\
\hline
000 \\
101 \\
\hline
1010 \\
101 \\
\hline
11110
\end{array}
$$

Thus, in general, we add the numbers $x2^{i-1}y_i$ $(i = 1, ..., m_2)$, using the notation of Section 4.3 except that now $xy = z$. In the ith summand x is multiplied by 2^{i-1} or by 0 according as y_i is 1 or 0, i.e. the number x is shifted by $i-1$ digits or replaced by zero according as y_i is 1 or 0. The additions are performed by incorporating an adder in the multiplier, and other electronic devices are incorporated to control the operation of the adder.

In more refined forms of multiplier the addition of zero summands may be eliminated and use is made of the representation of numbers in the forms

$$\sum_{i=1}^{n} 2^{i-1} x_i$$

where

$$x_i \in \{1, 0, -1\}$$

We note that, if $m < n$,

$$\sum_{i=1}^{k} 2^{a_i} + \sum_{i=m+1}^{n} 2^i + \sum_{i=0}^{m-1} 2^i = \sum_{i=1}^{k} 2^{a_i} + 2^{n+1} - 2^m - 1$$

so that, if $a_1, ..., a_k$ are integers and

$$a_1, ..., a_k > n$$

then a binary number whose first (in increasing order of significance) $n+1$ digits are all equal to 1 except the mth and whose remaining digits are 0 except the $(1 + a_1), ..., (1 + a_k)$th may be written

$$\sum_{i=1}^{k} 2^{a_i} + 2^{n+1} - 2^m - 2^0$$

The multiplication process may then involve subtraction as well as addition and subtraction of zero may be omitted as for addition (see above). For example, the evaluation of the product

$$1110111101.1001011111$$

would, according to the additive method, involve consideration of the expression

$$1110111101(2^9+2^6+2^4+2^3+2^2+2^1+2^0)$$

but we may, instead, consider the expression

$$1110111101(2^9+2^7-2^5-2^0)$$

which yields the following evaluation:

		Product of 1110111101 by
−	1110111101	2^0 (subtracted)
−	1110111101	2^5 (subtracted)
−	111101101011101	2^0+2^5 (subtracted)
+	1110111101	2^7 (added)
+	10110001100100011	
+	1110111101	2^9 (added)
	10001101110100100011	Final product

Examples 4*D*

1. Rewrite the following products in binary notation and then evaluate them by conventional methods.
   (i) 47.169   (ii) 38.143   (iii) 54.92.
2. Evaluate, by the addition and subtraction methods, the products
   (i) 1110100011011.100001111011
   (ii) 11011100010101.10100001111101

Solutions 4*D*

1. (i) 47.169 = 101111.10101001

```
        101111
      10101001
      _____
        101111
        000000
        000000
       101111
       000000
       101111
       000000
       101111
      _____
    1111100000111
```

2. (i) Consider

$$1110100011011(2^{11}+2^7-2^2-2^0)$$

$$
\begin{array}{ll}
- & 1110100011011 \\
- & 1110100011011 \\
\hline
- & 1001000110000111 \\
+ & 1110100011011 \\
\hline
+ & 11011111101111111001 \\
+ & 1110100011011 \\
\hline
& 111101101101001111111001
\end{array}
$$

4.8 The use of 3-input decision elements as flip-flops*

A flip-flop is a device which in its simplest form may be likened to a decision element which has two inputs and one output. The inputs (from top to bottom in the diagrams) are known as 'set' or 'trigger' and 'reset' respectively. Initially the states of all three wires correspond to the truth value F (Figure 4.6(a)). If the state of the set wire is then made to correspond to the truth value T the state of the output corresponds to T (Figure 4.6(b)),

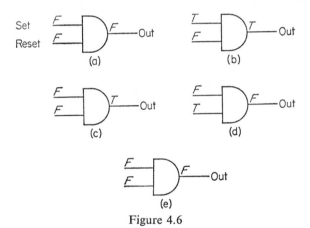

Figure 4.6

and remains so, even if the state of the set wire then reverts to that corresponding to F (Figure 4.6(c)). Thus the state of the output wire corresponds to the truth value of the proposition, 'The state of the set wire corresponds, or has

---

* The term 'flip-flop' is commonly used in place of the more exact term 'bistable' which is its meaning here.

corresponded, to the truth value *T'*, and a flip-flop may be regarded as a single-digit memory device. Ultimately, of course, the need to remember in this particular way will have passed, so it must be possible to arrange for the whole flip-flop to revert to its original state. If the state of the reset wire is now made to correspond to the truth value *T*, the state of the output wire reverts to that corresponding to the truth value *F* (Figure 4.6(d)) and remains so when the state of the reset wire reverts to that corresponding to the truth value *F* (Figure 4.6(e), which is identical with Figure 4.6(a)).

Since the state of the output wire at a given time may depend on its state at an earlier time, it seems plausible to suppose that a flip-flop may be made from a 3-input decision element by connecting the output to one of the inputs.

Let us denote by P, Q, R, formulae corresponding to the inputs called fed-back output, set and reset respectively, and by $\Phi(\,,\,)$ the ternary functor corresponding to the decision element (cf. Figure 4.7). We shall now determine which ternary functors,[20] if any, correspond to decision elements which can be used as flip-flops in this way. First we shall suppose that the functor $\Phi(\,,\,)$ can be so used and establish the necessity of six conditions concerning its truth table. The sufficiency of these conditions will then follow easily and

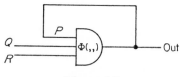

Figure 4.7

we shall thereby establish that there are exactly four ternary functors which are solutions to the problem. If P, Q, R take the truth values x, y, z respectively and, when *T* and *F* are renamed 0 and 1 respectively,

$$w = x + 2y + 4z + 1$$

we shall refer to the assignment of truth values to the arguments of the formula $\Phi(P, Q, R)$ as assignment number w. Thus the possible values of w are $1, \ldots, 8$. The six conditions relate to the truth values of the formula $\Phi(P, Q, R)$ under assignments $3, \ldots, 8$ respectively.

Initially P, Q, R take the truth value *F* and the output corresponds to the truth value *F*, so that under assignment number 8, $\Phi(P, Q, R)$ takes the truth value *F* (cf. Figure 4.6(a)). As this is the first of the six conditions, this latter entry of *F* in the truth table shown below is followed by a figure 1 and similar numbers will be used in subsequent cases. The situation in Figure 4.6(b) then indicates that, under assignment number 6, the formula $\Phi(P, Q, R)$ takes the

Assignment number	P	Q	R	$\Phi(P, Q, R)$	
1	T	T	T	T or F	
2	F	T	T	T or F	
3	T	F	T	F	5
4	F	F	T	F	6
5	T	T	F	T	3
6	F	T	F	T	2
7	T	F	F	T	4
8	F	F	F	F	1, 7

truth value T (No. 2). The situation is now as shown in Figure 4.8, but is transient since the output has just changed and the input corresponding to P must follow and change to the state corresponding to the truth value F. As the situation illustrated in Figure 4.6(b) requires the output to remain in the state corresponding to the truth value T at least until the state of one of the set and reset wires is changed, the formula $\Phi(P, Q, R)$ must, under assignment

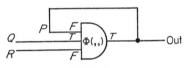

Figure 4.8

number 5, take the truth value T (No. 3). The situation illustrated in Figure 4.6(c) now requires that, under assignment number 7, the formula $\Phi(P, Q, R)$ takes the truth value T (No. 4). Use of the reset wire (cf. Figure 4.6(d)) now requires that, under assignment number 3, $\Phi(P, Q, R)$ takes the truth value F, (No. 5). This situation is transient, as is clear from Figure 4.9, so we require also that, under assignment number 4, $\Phi(P, Q, R)$

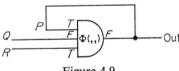

Figure 4.9

takes the truth value F (No. 6). The situation in Figure 4.6(e) now requires again that, under assignment number 8, $\Phi(P, Q, R)$ takes the truth value F (No. 7 as well as 1).

It is clear that, conversely, if the truth table of the functor $\Phi(, ,)$ contains these six entries then all the conditions for the simulation of a flip-flop are satisfied. Since there are exactly $4\,(=2^2)$ ways of making entries in the last column of the table for the assignments whose numbers are 1 and 2, there are exactly 4 suitable ternary functors.

Let us denote by $\Phi_z(\ ,\ ,\ )$ the functor in the first and second lines of whose truth table the right-hand entries are x, y respectively, where

$$x, y \in \{0(T), 1(F)\}$$

and

$$z = x + 2y + 1$$

As convenient methods of describing the truth tables of these four functors we note that

$$\Phi_1(P, Q, R) =_T (P \not\supset R) \vee Q$$

$$\Phi_2(P, Q, R) =_T [\sim R, P, Q]$$

$$\Phi_3(P, Q, R) =_T L_2(P, Q, \sim R)$$

$$\Phi_4(P, Q, R) =_T P \vee Q \not\supset R$$

4.9 Simple uses of flip-flops

A flip-flop may be used to decide, given a formula $\Phi(P_1, ..., P_n)$ (where the syntactical variables $P_1, ..., P_n$ denote exactly those propositional variables occurring in the formula) and a set $\mathscr{E}$ $(1 \leqslant \overline{\overline{\mathscr{E}}} \leqslant 2^n)$ of assignments of truth values to formulae $P_1, ..., P_m$ $(m \geqslant n)$, whether the formula $\Phi(P_1, ..., P_n)$ takes the truth value T under at least one assignment of $\mathscr{E}$. It is only necessary to set up a decision mechanism for the formula $\Phi(P_1, ..., P_n)$, connecting the output of this mechanism to the set wire of a flip-flop (leaving the reset wire in the state corresponding to the truth value F), and then supply n-tuples of inputs corresponding to those n-tuples of truth values which are members of $\mathscr{E}$. If the formula $\Phi(P_1, ..., P_n)$ takes the truth value T under at least one such assignment then the input from the decision mechanism to the flip-flop will correspond to the truth value T at least once. From that time onwards the output of the flip-flop will correspond to the truth value T, but it will not do so until then. Hence, if $\Phi(P_1, ..., P_n)$ takes the truth value F under all assignments of $\mathscr{E}$, the output of the flip-flop will remain in the state corresponding to the truth value F.

It may, for instance,[21] be necessary to decide whether, under all those assignments of a set $\mathscr{E}$ of truth values to the propositional variables $p_1, ..., p_m$ $(m \geqslant 2)$ under which p_1 takes the truth value T, p_2 takes the truth value T also. This is the case if and only if the formula

$$p_1 \not\supset p_2$$

does not take the truth value T under at least one assignment of $\mathscr{E}$ and the latter question is, as explained in the last paragraph, decidable.

As a further use of flip-flops let us consider the shift register mentioned at the end of Section 4.6. We must be able to make the output of the kth flip-flop after $i+1$ units of time have the same physical state as the previous (i.e. $(k-1)$th) flip-flop had after i units of time ($k = 2, 3, ..., n$; n being defined as at the end of Section 4.6). The units of time $i, i+1, ...$ are defined by control signals usually referred to as shift pulses. Thus receipt of a shift pulse causes all digits to shift one position, no action being caused in any other way.

Such a shift register may be constructed from flip-flops of a somewhat different type from those of the previous section. If the state of the output of the kth flip-flop after i units of time corresponds to the truth value of the formula P_{ik} ($i = 0, 1, ...$; $k = 1, 2 ...$) and Q_i takes the truth value T if and only if a shift pulse is received after i units of time then

$$P_{i+1,k} =_{\mathrm{T}} [P_{i,k-1}, Q_{i+1}, P_{ik}] \quad (i = 0, 1, ...; k = 1, 2, ...) \tag{A}$$

We therefore require, as is explained at the end of this paragraph, a flip-flop in which the roles of the set and reset inputs are to some extent unified. The state of the output after $i+1$ units of time must correspond to the same truth value as that to which the state of the first input corresponded after i units of time, provided that the second input, after $i+1$ units of time, is in a state corresponding to the truth value T. Otherwise the state of the output after $i+1$ units of time is to be identical with that after i units of time. Such a flip-flop may, if the first and second inputs correspond (at the appropriate times) to the formulae Q, R respectively,* be made from a conditioned disjunction decision element and a delay unit. We have only to feed back the output through a delay unit, making the resulting input correspond to the formula P and note that the output then corresponds to the formula $[Q, R, P]$ (see Figure 4.10). It then follows from (A) that a shift register may

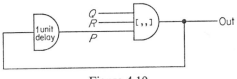

Figure 4.10

be made from these flip-flops by connecting the output of each (through the delay unit of the next) to the first input of the next, the second input being connected to the source of shift pulses (see Figure 4.11, in which the outputs

* A delay unit is, of course, needed for the input corresponding to Q (see below).

shown correspond to the cases where shift pulses are always received once per unit of time).

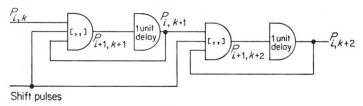

Figure 4.11

In order to check numerical work done by a computer, as a safeguard against failure of the computer hardware, numbers are often given an extra digit, more significant than any of the normal digits, in such a way that every number now has an even number of binary ones. Thus, for example, 7 and 17 would be written in binary form not as 111, 10001 respectively but as 1111, 010001. This additional digit is known as the parity digit. Thus, after transfer in or out of any store, a number can be checked in such a way that any single-digit error will be detected. Provided that inputs corresponding to all digits are available simultaneously (as in the case of certain shift registers) we may, to construct a *parity checker*, make use of the fact (cf. the Leśniewski–Mihailescu theorem, Section 3.3) that, if E^{n-1} denotes $n-1$ consecutive symbols E, then the formula $E^{n-1}P_1 \ldots P_n$ takes the truth value T if and only if an even number of the formulae $P_1, \ldots, P_n$ take the truth value F. Thus, if n is even (odd), the formula $E^{n-1}P_1 \ldots P_n$ takes the truth value T if and only if an even (odd) number of the formulae $P_1, \ldots, P_n$ take the truth value T.

Another cause of computer error is the difficulty of feeding into a digital computer information of a non-digital type which must be converted to binary numbers. When two consecutive numbers differ in respect of all their digits, borderline effects may (for practical reasons, beyond the scope of this book) cause errors in several digits. These difficulties may be avoided by representing integers as combinations of zeros and units in such a way that two consecutive integers differ with regard to only one digit. Such a representation is known as a *Gray code* and we shall consider the use of flip-flops in the construction of a Gray code to binary converter.

We may, if G_i represents, in the Gray code, the ith (in increasing order of significance) digit $(i = 1, \ldots, n)$ of the integer X $(0 \leqslant X \leqslant 2^n - 1)$ take $G_i = 0$ if and only if one of the congruences

$$X \equiv 0, \ldots, 2^{i-1}-1 \quad 2^i+2^{i-1}, \ldots, 2^{i+1}-1 \quad (\mathrm{mod}\, 2^{i+1}) \qquad (A)$$

holds. An increase of 1 in the value of X will change the truth value of the above disjunction of congruences for only one value of i, since a change occurs if and only if $x+1$ is divisible by 2^{i-1} but not by 2^i.

Let B_i represent the ith digit of X (in increasing order of significance) in the ordinary binary notation ($i = 1, ..., n$). For the particular Gray code discussed above we shall show that

$$B_n = G_n \tag{1}$$

and that, for $i = 1, ..., n-1$

$$B_i = B_{i+1} \quad \text{if and only if } G_i = 0 \tag{2}$$

We note that, by (A),

$$G_n = 0 \quad \text{if and only if} \quad X \leqslant 2^{n-1} - 1$$

that is,

$$\text{if and only if} \quad B_n = 0$$

so that (1) follows at once. We note also that $B_i = B_{i+1} = 0$ if and only if one of the congruences

$$X \equiv 0, ..., 2^{i-1} - 1 \pmod{2^{i+1}}$$

holds and that $B_i = B_{i+1} = 1$ if and only if one of the congruences

$$X \equiv 2^i + 2^{i-1}, ..., 2^{i+1} - 1 \pmod{2^{i+1}}$$

holds ($i = 1, ..., n-1$). Thus (2) follows at once from (A).

Hence, if the formulae $P_1, ..., P_n, Q_1, ..., Q_n$ are related, in the usual way, to the digits $G_1, ..., G_n, B_1, ..., B_n$ respectively,

$$Q_n =_{\text{T}} P_n$$

and

$$Q_i =_{\text{T}} Q_{i+1} \not\equiv P_i =_{\text{T}} [\sim P_i, Q_{i+1}, P_i] \quad (i = 1, ..., n-1)$$

Thus, with the notation of Section 4.7,

$$Q_i =_{\text{T}} \Phi_2(P_i, Q_{i+1}, P_i) \quad (i = 1, ..., n-1)$$

and a Gray code to binary converter may be constructed from a flip-flop (provided that it corresponds to the functor $\Phi_2(\ ,\ ,\ )$) if the Gray code digit inputs are supplied serially in decreasing order of significance. Such a converter is of considerable practical value, since by far the majority of digital computers are constructed to perform calculations on numbers in the ordinary binary code, rather than in a Gray code.

Examples 4E

1. Prove that the universal decision element of Section 3.2 may be used as a flip-flop.

2. Prove that if

$$\Phi(P, Q, R, S, U) =_\mathrm{T} [P \not\Rightarrow Q, R, S \equiv U]$$

then a decision element for the functor $\Phi(\,,\,,\,,\,)$ may be used as a flip-flop. Prove also that if

$$\Psi(P, Q, R, S) =_\mathrm{T} P \equiv [Q, R, S]$$

then the result fails for the functor $\Psi(\,,\,,\,)$.

3. Show how to use a flip-flop to determine whether, under all members of a set $\mathscr{E}$ of assignments of truth values to the propositional variables p, q, r, s,

  (i) the truth values of p and q are the same,

  (ii) the truth value of the formula $p \vee q$ differs from those of both r and s.

Solutions 4E

1. $(P \not\Rightarrow R) \vee Q =_\mathrm{T} [t, (P \vee f) \mathbin{\&} (R/t), Q]$

3. (i) $p =_\mathrm{T} q$ under all assignments of $\mathscr{E}$ if and only if the formula $p \not\equiv q$ does not take the truth value T under any assignment of $\mathscr{E}$.

Chapter 5

Designation Numbers and Their Applications

5.1 Definition of designation numbers and simple examples of their uses

We have seen in Chapter 1 that given a formula $\Phi(P_1, ..., P_n)$ containing exactly the n distinct propositional variables denoted by the syntactical variables $P_1, ..., P_n$, we may determine the truth value of $\Phi(P_1, ..., P_n)$ when the truth values of $P_1, ..., P_n$ are known. Since there are exactly 2^n ways of assigning truth values to these n variables there are exactly 2^n such determinations. The results of these 2^n determinations may be set out as a sequence in any of $(2^n)!$ ways and we may refer to any of these $(2^n)!$ sequences as a *designation sequence* of the formula $\Phi(P_1, ..., P_n)$. (It is sometimes, as we shall see below, useful to consider such a sequence of 2^n truth values when one or more of the n propositional variables is absent from the formula considered.) It often happens that, instead of formulae of the propositional calculus, expressions of Boolean algebra (cf. Section 1.10) are considered and, correspondingly, the variables $A_1, ..., A_n$ of this algebra are assigned the values 1 and 0 corresponding to the assignments of the truth values T and F respectively to the related propositional variables. (We shall, in future, use 1 and 0 in place of I and O respectively.) When a designation sequence is so replaced by a sequence of digits a binary number is, of course, formed and this number is known as a *designation number* of the corresponding expression of Boolean algebra.

We have not, so far, attempted to make a definite choice of one way (from the $(2^n)!$ possible ways) of ordering the 2^n entries in a designation sequence or designation number. In future we shall take as the ith element ($1 \leqslant i \leqslant 2^n$, the first element occurring on the left), that corresponding to the assignment of the values $x_1, ..., x_n$ to $A_1, ..., A_n$ respectively, where

$$i = 1 + \sum_{j=1}^{n} 2^{j-1} x_j$$

Thus the successor of the binary number formed by the values of $A_1, ..., A_n$ is the number of the position of the corresponding value of the Boolean expression in the designation number.

Thus, for example, if $n = 3$ and the values of $P_1, P_2, P_3, \Phi(P_1, P_2, P_3)$ are

$$p \quad q \quad r \quad (p \lor q)\,\&\sim(r \lor \sim p)$$

respectively, then the corresponding Boolean expression is

$$(A_1 \cup A_2) \cap (A_3 \cup A_1')'$$

and the designation numbers of the relevant expressions are as follows. The eight (decimal) numbers in the first row are the successors of the binary numbers formed by the first three digits in the corresponding columns and thus the position numbers of the digits of the designation numbers.

	1	2	3	4	5	6	7	8
A_1	0	1	0	1	0	1	0	1
A_2	0	0	1	1	0	0	1	1
A_3	0	0	0	0	1	1	1	1
$A_1 \cup A_2$	0	1	1	1	0	1	1	1
A_1'	1	0	1	0	1	0	1	0
$A_3 \cup A_1'$	1	0	1	0	1	1	1	1
$(A_3 \cup A_1')'$	0	1	0	1	0	0	0	0
$(A_1 \cup A_2) \cap (A_3 \cup A_1')'$	0	1	0	1	0	0	0	0

Alternatively we could work in the propositional calculus using designation sequences. Let us, for example, consider the case when $n = 4$ and the values of $P_1, P_2, P_3, P_4, \Phi(P_1, P_2, P_3, P_4)$ are

$$p \quad q \quad r \quad s \quad (p \lor \sim q)\,\&\sim(r\,\&\,s)$$

respectively.

	1	2	3	4	5	6	7	8	9	10	11	12	13	14	15	16
p	F	T	F	T	F	T	F	T	F	T	F	T	F	T	F	T
q	F	F	T	T	F	F	T	T	F	F	T	T	F	F	T	T
r	F	F	F	F	T	T	T	T	F	F	F	F	T	T	T	T
s	F	F	F	F	F	F	F	F	T	T	T	T	T	T	T	T
$\sim q$	T	T	F	F	T	T	F	F	T	T	F	F	T	T	F	F
$p \lor \sim q$	T	T	F	T	T	T	F	T	T	T	F	T	T	T	F	T
$r\,\&\,s$	F	F	F	F	F	F	F	F	F	F	F	F	T	T	T	T
$\sim(r\,\&\,s)$	T	T	T	T	T	T	T	T	T	T	T	T	F	F	F	F
$(p \lor \sim q)\,\&\sim(r\,\&\,s)$	T	T	F	T	T	T	F	T	T	T	F	T	F	F	F	F

As a simple application of designation numbers we shall consider a new method of reduction to disjunctive normal form. We note first that if

$$i = 1 + \sum_{j=1}^{n} 2^{j-1} x_j$$

and all digits except the ith of the designation number of an expression built out of exactly the n variables $A_1, ..., A_n$ are zero, the ith digit being 1, then the Boolean expression is equal to a product determined as follows. Let the distinct values of j ($1 \leqslant j \leqslant n$) for which $x_j = 1$ be

$$j_{i1}, ..., j_{ik}$$

and let the remaining values of j be

$$j_{i,k+1}, ..., j_{in}$$

Then the required product is

$$A_{j_{i1}} \cap A_{j_{i2}} \cap ... \cap A_{j_{ik}} \cap A_{j_{i,k+1}}{}' \cap A_{j_{i,k+2}}{}' \cap ... \cap A_{j_{in}}{}'$$

Clearly if (corresponding to the ith digit of the designation number)

$$A_{j_{i1}}, ..., A_{j_{ik}} = 1 \qquad A_{j_{i,k+1}}, ..., A_{j_{in}} = 0$$

then the above expression is equal to 1 and, in the remaining $2^n - 1$ cases, it is equal to 0. Let us denote this expression by

$$\Phi_i(A_i, ..., A_n) \quad (i = 1, ..., 2^n)$$

Then, more generally, if the designation number of an expression built out of the above n variables is such that the position numbers of its unit digits are $N_1, ..., N_\alpha$ then a disjunctive normal form is given by

$$\Phi_{N_1}(A_1, ..., A_n) \cup \Phi_{N_\alpha}(A_1, ..., A_n) \cup ... \cup \Phi_{N_\alpha}(A_1, ..., A_n)$$

since the above expression is equal to 1 if and only if at least one of the expressions $\Phi_{N_\beta}(A_1, ..., A_n)$ ($1 \leqslant \beta \leqslant \alpha$) is equal to 1. (More precisely, this disjunctive normal form is a Boolean representative of a disjunctive normal form of the propositional calculus formula corresponding to the original Boolean expression.)

For example, the expression

$$(A_1 \cup A_2) \cap (A_3 \cup A_1{}')'$$

considered above was found to have 01010000 as its designation number. Considering, as above, the designation number 01000000 we find the unit digit corresponding to the assignment where

$$A_1 = 1 \qquad A_2 = 0 \qquad A_3 = 0$$

and we therefore obtain the corresponding expression

$$A_1 \cap A_2{}' \cap A_3{}'$$

The other unit digit in the designation number is dealt with similarly and we obtain as a disjunctive normal form the expression

$$(A_1 \cap A_2{}' \cap A_3{}') \cup (A_1 \cap A_2 \cap A_3{}')$$

5

A slight extension of the method enables us to consider conjunctive normal forms. Having obtained the designation sequence of a formula we may obtain immediately the designation sequence for its negation. We may then, proceeding as above, determine a disjunctive normal form for this negation. The original formula will then have the same truth table as the negation of this disjunctive normal form, so that a conjunctive normal form of the original formula may be obtained, using the de Morgan laws and the equation $\sim \sim P =_T P$.

Let us, for example, consider the formula

$$(p \,\&\, q \vee r) \,\&\, \sim [\![(q \vee \sim s) \,\&\, p]\!]$$

of the propositional calculus. It is easily seen that its designation sequence is

$$FFFFTFTFFFFFTTTF$$

Thus the designation sequence of its negation is

$$TTTTFTFTTTTTFFFT$$

and, proceeding as in the previous example, and noting that all eight elements of the latter sequence corresponding to the assignment of the truth value F to r are T, a disjunctive normal form of the negation is found to be

$$\sim r \vee p \,\&\, \sim q \,\&\, r \,\&\, \sim s \vee p \,\&\, q \,\&\, r \,\&\, \sim s \vee p \,\&\, q \,\&\, r \,\&\, s$$

which simplifies easily (by the distributive law, cf. Chapter 1) to the formula

$$\sim r \vee p \,\&\, r \,\&\, \sim s \vee p \,\&\, q \,\&\, r \,\&\, s$$

Thus the given formula takes the same truth value as the formula

$$\sim (\sim r \vee p \,\&\, r \,\&\, \sim s \vee p \,\&\, q \,\&\, r \,\&\, s)$$

or as the conjunctive normal form

$$r \,\&\, (\sim p \vee \sim r \vee s) \,\&\, (\sim p \vee \sim q \vee \sim r \vee \sim s)$$

since, by the de Morgan laws,

$$\sim \sum_{i=1}^{\alpha} \prod_{j=1}^{\beta_i} P_{ij} =_T \prod_{i=1}^{\alpha} \sum_{j=1}^{\beta_i} \sim P_{ij} \quad (\beta_1 = 1, 2, \ldots; \ \ldots; \beta_\alpha = 1, 2, \ldots; \alpha = 1, 2, \ldots)$$

and $\sim \sim P =_T P$. In general, to obtain a conjunctive normal form from the negation of a disjunctive normal form we have only to interchange all conjunction and disjunction functors, remove the existing negation symbols and negate all previously unnegated occurrences of propositional variables.

In the last example considered above we modified the previously discussed method, selecting the formula $\sim r$ corresponding to eight assignments, instead of eight separate conjunctions. Similarly, rather than use the distributive law, we could have selected the formula $p \& r \& \sim s$ corresponding to two assignments since the entries in positions 6 and 8 of the designation sequence (corresponding to the two cases where p, r, s take the truth values T, T, F respectively) were both T.

More generally, if $i_1, \ldots, i_k$ are distinct integers of the set $\{1, \ldots, n\}$ and, for some assignment of truth values to $P_{i_1}, \ldots, P_{i_k}$, the entry in the designation sequence is independent of the truth values of the remaining $n-k$ propositional variables, we may take a single conjunction corresponding to the 2^{n-k} cases. If, under these 2^{n-k} assignments, P_{i_l} takes the truth value x_{i_l} $(l = 1, \ldots, k)$ and Q_l is P_{i_l} or $\sim P_{i_l}$ according as x_{i_l} is T or F then our conjunction is

$$\prod_{l=1}^{k} Q_{i_l}$$

If $y_{i_l} = 1$ or 0 according as x_{i_l} is T or F, the corresponding 2^{n-k} designation numbers will be the numbers

$$1 + \sum_{l=1}^{k} 2^{i_l-1} y_{i_l} + \sum_{j=k+1}^{n} 2^{i_j-1} y_{i_j} \quad [\![y_{i_{k+1}}, \ldots, y_{i_n} \in \{0, 1\}]\!]$$

where

$$\{i_{k+1}, \ldots, i_n\} = \{1, \ldots, n\} - \{i_1, \ldots, i_k\}$$

Thus if, for some choice of values for k, $i_1, \ldots, i_k$, $y_{i_1}, \ldots, y_{i_k}$ the 2^{n-k} digits whose position numbers are

$$1 + \sum_{l=1}^{k} 2^{i_l-1} y_{i_l} + \sum_{j=k+1}^{n} 2^{i_j-1} y_{i_j}$$

are all equal, we can replace 2^k disjuncts in the disjunctive normal form by a single disjunct. In the above example we considered, for $n = 4$, the two cases

$$k = 1 \quad i_i = 3 \quad y_3 = 0$$

$$k = 3 \quad i_1 = 1 \quad i_2 = 3 \quad i_3 = 4 \quad y_1 = 1 \quad y_3 = 1 \quad y_4 = 0$$

Examples 5A

1. Find the designation numbers of
   (i) $(A_1 \cup A_2)' \cap [\![A_3' \cup (A_1' \cap A_2)]\!]'$
   (ii) $\{A_1 \cap [\![A_2' \cup (A_3 \cap A_4')']\!]'\} \cap (A_1' \cap A_3)'$
2. Find the designation sequences of
   (i) $(p \supset q \vee r) \& [\![q \supset \sim (p \& r)]\!]$
   (ii) $p \& q \equiv \sim (r \vee \sim s)$
3. Reduce to both normal forms
   (i) $q \& (p \not\equiv r)$
   (ii) $(p \equiv q) \vee (r \equiv s) \supset \sim [\![p \& (q \vee \sim s)]\!]$

4. Find Boolean expressions whose designation numbers are

(i)   1110101110101011

(ii)   1100001111000011

Solutions 5A

1. (i)

	1	2	3	4	5	6	7	8
A_1	0	1	0	1	0	1	0	1
A_2	0	0	1	1	0	0	1	1
A_3	0	0	0	0	1	1	1	1
$A_1 \cup A_2$	0	1	1	1	0	1	1	1
$(A_1 \cup A_2)'$	1	0	0	0	1	0	0	0
A_3'	1	1	1	1	0	0	0	0
A_1'	1	0	1	0	1	0	1	0
$A_1' \cap A_2$	0	0	1	0	0	0	1	0
$A_3' \cup (A_1' \cap A_2)$	1	1	1	1	0	0	1	0
$[\![A_3' \cup (A_1' \cap A_2)]\!]'$	0	0	0	0	1	1	0	1
$(A_1 \cup A_2)' \cap [\![A_3' \cup (A_1' \cap A_2)]\!]'$	0	0	0	0	1	0	0	0

2. (i)

	1	2	3	4	5	6	7	8
p	F	T	F	T	F	T	F	T
q	F	F	T	T	F	F	T	T
r	F	F	F	F	T	T	T	T
$q \vee r$	F	F	T	T	T	T	T	T
$p \supset q \vee r$	T	F	T	T	T	T	T	T
$p \,\&\, r$	F	F	F	F	F	T	F	T
$\sim(p \,\&\, r)$	T	T	T	T	T	F	T	F
$q \supset \sim(p \,\&\, r)$	T	T	T	T	T	T	T	F
$(p \supset q \vee r) \,\&\, [\![q \supset \sim(p \,\&\, r)]\!]$	T	F	T	T	T	T	T	F

3. (i) The given formula has the designation sequence

FFFTFFTF

Hence the negation of the given formula has the designation sequence

TTTFTTFT

and this latter formula has, as a disjunctive normal form, the formula

$\sim q \vee \sim p \,\&\, q \,\&\, \sim r \vee p \,\&\, q \,\&\, r$

Hence the given formula has the same truth table as the formula

$\sim(\sim q \vee \sim p \,\&\, q \,\&\, \sim r \vee p \,\&\, q \,\&\, r)$

and it has, as a conjunctive normal form, the formula

$q \,\&\, (p \vee \sim q \vee r) \,\&\, (\sim p \vee \sim q \vee \sim r)$

4. (i) The entries whose numbers are 1, 3, 5, 7, 9, 11, 13, 15 are all 1, giving rise to a summand (regarding ∪ as addition) of A_1'. The other unit entries have numbers 2, 8, 16, giving rise to summands of $A_1 \cap A_2' \cap A_3' \cap A_4'$ (2) and $A_1 \cap A_2 \cap A_3$ (8, 16). Thus a suitable choice for the required expression is

$$A_1' \cup (A_1 \cap A_2' \cap A_3' \cap A_4') \cup (A_1 \cap A_2 \cap A_3)$$

5.2 Application to the determination of simplest normal form[9, 22]

We shall now use designation numbers or sequences to determine, given an arbitrary formula of the propositional calculus, the disjunctive normal form of the formula such that the two following conditions are satisfied. (The formula obtained may not be unique.)

(i) No disjunct may be replaced by a conjunction of some (but not all) of the present conjuncts of that disjunct (repeated occurrences of a disjunct being treated as distinct disjuncts).

(ii) The number of disjuncts is the least possible.

An alternative approach is by means of Veitch diagrams. This method is discussed by Phister,[23] and a further alternative has been discussed by J. P. Roth.[24]

Let the given formula contain exactly n distinct propositional variables, which we shall denote by the syntactical variables $P_1, ..., P_n$. A formula

$$\prod_{i=1}^{k} Q_i \quad [\![Q_1, ..., Q_k \in \{P_1, ..., P_n, \sim P_1, ..., \sim P_n\}]\!]$$

is said to be a *prime implicant* of the given formula if it satisfies the conditions (iii), (iv) given below.

(iii) The formula

$$\prod_{i=1}^{k} Q_i$$

takes the truth value T only when the given formula does so.

(iv) If $l < k$ and

$$R_1, ..., R_l \in \{Q_1, ..., Q_k\}$$

then the formula $\prod_{i=1}^{l} R_i$ and the given formula are capable of taking the respective truth values T, F simultaneously. Clearly, by (i), the simplest normal form must be a disjunction of prime implicants. We therefore determine all the prime implicants of the given formula and then, by considering the designation sequences corresponding to these prime implicants, determine which disjunction(s) of prime implicants satisfy(ies) (ii).

Let us, for example, consider the formula

$$(p \,\&\, q \lor r) \,\&\sim [\![(q \lor \sim s) \,\&\, p]\!]$$

whose designation sequence was found (cf. Section 5.1, the example of conjunctive normal form) to be

$$FFFFTFTFFFFFTTTF$$

The position numbers of the T's in this sequence are

$$5, 7, 13, 14, 15$$

Clearly the numbers 5, 7, 13, 15 correspond to the conjunction

$$\sim p \,\&\, r$$

Since the conjuncts

$$\sim p \quad \text{(cf. position number 1)}$$

$$r \quad \text{(cf. position number 6)}$$

are not prime implicants, the formula

$$\sim p \,\&\, r$$

is a prime implicant.

To the number 14 there corresponds the conjunction

$$p \,\&\sim q \,\&\, r \,\&\, s$$

so we must consider the applicability of (iv) to the formulae

$$\sim q \,\&\, r \,\&\, s \quad p \,\&\, r \,\&\, s \quad p \,\&\sim q \,\&\, s \quad p \,\&\sim q \,\&\, r$$

These correspond respectively to the pairs of position numbers shown below.

$$13, 14 \quad 14, 16 \quad 10, 14 \quad 6, 14$$

(obtained by adding to 14 the numbers $-2^0, 2^1, -2^2, -2^3$, the negative sign corresponding to the truth value of the removed variable being changed from T to F). Since 13 is listed in the original position numbers but 16, 10 and 6 are not, the only possible prime implicant obtained from 14 is $\sim q \,\&\, r \,\&\, s$.

Since any conjunction involving the formulae $\sim p, r$ and at least one other conjunct is not a prime implicant the only other possible prime implicants corresponding to 5, 7, 13 and 15 are

$$\sim p \,\&\, q \,\&\, s \quad \sim p \,\&\, q \,\&\sim s \quad \sim p \,\&\sim q \,\&\, s \quad \sim p \,\&\sim q \,\&\sim s$$

$$q \,\&\, r \,\&\, s \quad \sim q \,\&\, r \,\&\, s \quad q \,\&\, r \,\&\sim s \quad \sim q \,\&\, r \,\&\sim s$$

and formulae obtained from these by deletion. However, considering, in the eight respective cases, the positions numbered 1, 3, 9, 11, 6, 8, 14, 16 in which the considered conjunction again takes the truth value T (subtracting 4 in the first four cases, corresponding to the change in the truth value of r and, similarly, adding 1 in the second four cases) we see that the only possible prime implicant is, in fact, the formula $\sim q \,\&\, r \,\&\, s$ of the previous paragraph.

As shown below, further deletions are impossible, the numbers in brackets being the relevant position numbers.

$$\sim q \,\&\, r \quad (6 = 14 - 8)$$
$$\sim q \,\&\, s \quad (10 = 14 - 4)$$
$$r \,\&\, s \quad (16 = 14 + 2)$$

We see, therefore, that the only two prime implicants are the formulae

$$\sim p \,\&\, r \quad \sim q \,\&\, r \,\&\, s$$

These correspond, respectively, to the positions numbered

$$5, 7, 13, 15, 13, 14$$

Although both prime implicants deal with position number 13, the first prime implicant is required for position number 5 and the second for position number 14. Thus neither prime implicant may be deleted and a simplest disjunctive normal form is the formula

$$\sim p \,\&\, r \vee \sim q \,\&\, r \,\&\, s$$

the only alternatives being trivial variants such as the formula

$$\sim q \,\&\, s \,\&\, r \vee r \,\&\, \sim p$$

Examples 5B

1. Find a simplest disjunctive normal form of the formula whose designation sequence is
$$TTFFFTFT$$
and which is built out of the propositional variables p, q, r.
2. Find simplest disjunctive and conjunctive normal forms of the formula
$$(p \supset q \vee r) \,\&\, (p \vee \sim q)$$

Solutions 5B

1. The positions numbered 1, 2, 6, 8 give rise to the formulae

$$\sim p \,\&\, \sim q \,\&\, \sim r \quad p \,\&\, \sim q \,\&\, \sim r \quad p \,\&\, \sim q \,\&\, r \quad p \,\&\, q \,\&\, r$$

By deletion we obtain for consideration the following formulae, reasons for failure (provisional success) being given in brackets by single (pairs of) numbers.

$$\sim p \,\&\, \sim q \,(1 + 4 = 5) \quad \sim p \,\&\, \sim r \,(3) \quad p \,\&\, \sim r \,(4) \quad \sim q \,\&\, r \,(5) \quad p \,\&\, q \,(4)$$
$$q \,\&\, r \,(7) \quad \sim q \,\&\, \sim r \,(1, 2) \quad p \,\&\, \sim q \,(2, 6) \quad p \,\&\, r \,(6, 8)$$

the latter three being prime implicants by the failure of

$$p\,(4) \quad \sim q\,(5) \quad r\,(5) \quad \sim r\,(3)$$

The three prime implicants

$$\sim q\,\&\sim r \quad p\,\&\sim q \quad p\,\&\,r$$

correspond, respectively, to the positions numbered

$$1,\,2 \quad 2,\,6 \quad 6,\,8$$

The prime implicant $\sim q\,\&\sim r$ must be used since position 1 is not dealt with otherwise, as must the prime implicant $p\,\&\,r$ (position 8). However, these two prime implicants deal with all positions (2, 6) dealt with by the remaining prime implicant. Thus a simplest disjunctive normal form is

$$\sim q\,\&\sim r \lor p\,\&\,r$$

2. (Conjunctive case.) We first obtain the designation sequence of the negation of the given formula and a simplest disjunctive normal form of this negation. The given formula has (as may easily be checked by the reader) the designation sequence

$$\textit{TFFTTTFT}$$

so the required designation sequence is

$$\textit{FTTFFFTF}$$

The positions numbered 2, 3, 7 give rise to the respective formulae

$$p\,\&\sim q\,\&\sim r \quad \sim p\,\&\,q\,\&\sim r \quad \sim p\,\&\,q\,\&\,r$$

Trivially, the formulae (obtained by deletion)

$$p\,\&\sim q\,(2+4 = 6) \quad p\,\&\sim r\,(4) \quad \sim q\,\&\sim r\,(1) \quad \sim p\,\&\sim r\,(1) \quad q\,\&\sim r\,(4)$$
$$\sim p\,\&\,r\,(5) \quad q\,\&\,r\,(8)$$

are not prime implicants, but the formula

$$\sim p\,\&\,q\,(3,\,7)$$

is a possible prime implicant. Since

$$\sim p\,(1) \quad q\,(4)$$

are not prime implicants, the only prime implicants are the formulae

$$p\,\&\sim q\,\&\sim r \quad \sim p\,\&\,q$$

A simplest disjunctive normal form is the formula

$$p\,\&\sim q\,\&\sim r \lor \sim p\,\&\,q$$

since the disjuncts correspond to the positions numbered 2; 3, 7 respectively. Hence a simplest conjunctive normal form of the given formula is the formula

$$(\sim p \lor q \lor r)\,\&\,(p \lor \sim q)$$

5.3 Application to the testing for logical dependence

If $P_1, ..., P_n$ denote mixed formulae such that no propositional variable occurs in more than one of these n formulae then it is clear that, by suitable assignments of truth values to the propositional variables occurring in $P_1, ..., P_n$, we may assign truth values to the formulae $P_1, ..., P_n$ in 2^n ways. If, on the other hand, the members of the set $\mathscr{E}$ are those propositional variables which occur in at least one of the formulae $P_1, ..., P_n$ and

$$\overline{\overline{\mathscr{E}}} = k < n$$

then it is impossible to assign truth values to the formulae $P_1, ..., P_n$ in 2^n distinct ways since it is impossible to assign truth values to the propositional variables occurring in them in more than 2^k ways.

In general, however, it is not immediately obvious whether truth values may be assigned to $P_1, ..., P_n$ in 2^n different ways. If truth values may be so assigned then the formulae $P_1, ..., P_n$ are said to be *logically independent*. Otherwise they are said to be *logically dependent*.

In order to test a (finite) set of formulae for logical dependence it is sufficient to consider their designation sequences or the corresponding designation numbers. Let us suppose, with the notation of the first paragraph of this section, that

$$\overline{\overline{\mathscr{E}}} = k \geqslant n$$

and let us construct designation sequences for all the formulae $P_1, ..., P_n$ with respect to the members of $\mathscr{E}$. If, to each of the 2^n ordered n-tuples of truth values of $P_1, ..., P_n$, there corresponds an integer i (whose value depends on the choice of n-tuple) such that $1 \leqslant i \leqslant 2^k$, and the ith truth values in the designation sequences of $P_1, ..., P_n$ form this n-tuple then, clearly, the formulae are independent and conversely.

Let us, for example, consider the three formulae

$$p \vee q \supset r \quad p \equiv q \,\&\sim (p \,\&\, s) \quad p \vee (r \nparallel s \,\&\sim q)$$

With respect to the propositional variables p, q, r, s (in that order) the designation sequences of these formulae correspond to designation numbers as shown below.

$p \vee q \supset r$	1 0 0 0 1 1 1 1 1 0 0 0 1 1 1 1
$p \equiv q \,\&\sim (p \,\&\, s)$	1 0 0 1 1 0 0 1 1 0 0 0 1 0 0 0
$p \vee (r \nparallel s \,\&\sim q)$	0 1 0 1 1 1 1 1 0 1 0 1 0 1 1 1

3 4 0 6 7 5 5 7 3 4 0 4 3 5 5 5

For convenience, the binary numbers formed by the digits of the columns are given below the columns (as decimal numbers). Since the number 2 never so occurs, the three formulae cannot take the truth values F, T, F respectively and they are therefore logically dependent.

On the other hand, the formulae

$$(p \vee q) \& r \quad p \supset \sim(q \equiv p \& r)$$

are logically independent. Proceeding as in the previous paragraph we obtain the following results.

$$
\begin{array}{llllllllll}
(p \vee q) \& r & 0 & 0 & 0 & 0 & 0 & 1 & 1 & 1 \\
p \supset \sim(q \equiv p \& r) & 1 & 0 & 1 & 1 & 1 & 1 & 1 & 0 \\
\hline
 & 2 & 0 & 2 & 2 & 2 & 3 & 3 & 1
\end{array}
$$

Since all four ($= 2^2$) integers 0, 1, 2, 3 occur on the last line, all ordered pairs of truth values are taken simultaneously by the formulae, which are therefore logically independent.

Examples 5C

1. Prove that the formulae

$$(p \supset q) \equiv p \& r \quad (p \equiv q) \& (p \equiv r)$$

   are logically independent and that every set of three formulae, consisting of the two above and a third containing no propositional variables other than p, q, r, is logically dependent.
2. Prove that the formulae

$$[r, p, q] \quad [p, r, q]$$

   are logically independent and find a third formula, containing no propositional variables other than p, q, r, such that the three formulae are logically independent.
3. Prove that the formula $p \supset q \& r$ cannot belong to a set of three logically independent formulae unless at least one formula of the set contains at least one propositional variable other than p, q, r.
4. Test the following sets of formulae for logical dependence. (Bracket notation for sets is omitted.)

   (i) $p \& q \supset r \vee s \quad r \supset (p \equiv q)$
   (ii) $\sim(p \& \sim r) \vee q \quad p \& (r \vee s) \quad q \equiv r \vee s$
   (iii) $p \& q \quad q \& r \quad p \& (r \supset s)$
   (iv) $p \equiv q \quad [p, r, \sim q] \quad p$

Solutions 5C

1.
$$
\begin{array}{llllllllll}
(p \supset q) \equiv p \& r & 0 & 1 & 0 & 0 & 0 & 0 & 0 & 1 \\
(p \equiv q) \& (p \equiv r) & 1 & 0 & 0 & 0 & 0 & 0 & 0 & 1 \\
\hline
 & 2 & 1 & 0 & 0 & 0 & 0 & 0 & 3
\end{array}
$$

Since the formulae can take the truth values F, T respectively only once (as 2 occurs only once in the last row), these two formulae and a third formula cannot take, under two different assignments, the truth values F, T, F; F, T, T respectively.

2.
$$
\begin{array}{llllllll}
[r,p,q] & 0 & 0 & 1 & 0 & 0 & 1 & 1 & 1 \\
[p,r,q] & 0 & 1 & 0 & 1 & 0 & 0 & 1 & 1 \\
\hline
 & 0 & 2 & 1 & 2 & 0 & 1 & 3 & 3
\end{array}
$$

We must add a third row, after the first two above, in such a way that two columns with the same decimal number always have different elements. A suitable new designation number is thus

$$00011110$$

(since it contains unequal digits in the four pairs of positions numbered $1, 5; 2, 4; 3, 6; 7, 8$) and this corresponds to the formula

$$p \mathbin{\&} q \not\equiv r$$

5.4 Application to constraints in circuit design

Under certain circumstances a decision mechanism for an n-variable formula may need to be used for some, rather than all, the 2^n n-tuples of inputs. In these circumstances the decision mechanism needs only to correspond to a formula taking the required truth value under circumstances corresponding to those n-tuples of inputs which actually occur. For instance, if p, q, r never take the truth values F, T, T respectively then

$$(p \equiv q) \equiv r =_{\mathrm{T}} [p \supset q, r, p \not\equiv q]$$

If, further, p, q, r never take the truth values T, T, F respectively then

$$(p \equiv q) \equiv r =_{\mathrm{T}} (p \not\equiv r) \vee q$$

so that, provided appropriate constraints apply, a decision mechanism for the formula $(p \equiv q) \equiv r$ may be replaced by decision mechanisms for the formulae

$$[p \supset q, r, p \not\equiv q] \quad (p \not\equiv r) \vee q.$$

A method of testing the validity of such replacements is to inspect those elements of the designation sequences which correspond to the relevant assignments. In the above example, where p, q, r never take the respective truth values F, T, T or T, T, F, the designation sequences and their relevant subsequences are shown below.

$$
\begin{array}{llll}
(p \equiv q) \equiv r & FTTFTFFT & FTT\ TF\ T \\
(p \not\equiv r) \vee q & FTTTTFTT & FTT\ TF\ T
\end{array}
$$

Reductions to simplest normal form may be carried out in a similar way, using the relevant parts of the designation sequences. For example, the formula $(p \equiv q) \equiv r$ considered above, subject to the same two constraints, has the designation sequence

$$FTT?TF?T$$

where a question mark indicates an irrelevant assignment of truth values and the designation sequence may be deemed to contain either truth value. (The symbol 'd', meaning 'don't care', is used by many authors in place of the question mark.) Corresponding to the truth values whose position numbers are 2, 3, 5, 8 we consider the formulae

$$p \& {\sim}q \& {\sim}r \quad {\sim}p \& q \& {\sim}r \quad {\sim}p \& {\sim}q \& r \quad p \& q \& r$$

and we may, if it will shorten any conjunction, consider the formulae

$$p \& q \& {\sim}r \quad {\sim}p \& q \& r$$

corresponding to the truth values whose position numbers are 4 and 7.
Since

$$2 + 2^1 = 4 \quad 3 + 2^2 = 7 \quad 5 + 2^1 = 7 \quad 8 - 2^0 = 7$$

we may, by arguments similar to those used in Section 5.3, conclude that none of the four initial conjunctions is a prime implicant. We then consider conjunctions obtained by deletion of one conjunct.

$p \& {\sim}q$	Reject (6)
$p \& {\sim}r$	Consider further (2, 4)
${\sim}q \& {\sim}r$	Reject (1)
${\sim}p \& q$	Consider further (3, 7)
${\sim}p \& {\sim}r$	Reject (1)
$q \& {\sim}r$	Consider further (3, 4)
${\sim}p \& {\sim}q$	Reject (1)
${\sim}p \& r$	Consider further (5, 7)
${\sim}q \& r$	Reject (6)
$p \& q$	Consider further (8, 4)
$p \& r$	Reject (6)
$q \& r$	Consider further (8, 7)

We must now consider, for possible deletion of conjuncts, the conjunctions

$$p \& {\sim}r \quad {\sim}p \& q \quad q \& {\sim}r \quad {\sim}p \& r \quad p \& q \quad q \& r$$

To this end we consider

$$p \quad {\sim}p \quad q \quad r \quad {\sim}r$$

and reject p (6), $\sim p$ (1), r (6), $\sim r$ (1), accepting q (3, 4, 7, 8). Thus the prime implicants are

$$p \,\&\sim r \quad q \quad \sim p \,\&\, r$$

and these deal with the positions numbered

$$2, 4 \quad 3, 4, 7, 8 \quad 5, 7$$

respectively. We must select prime implicants to deal with the positions numbered 2, 3, 5, 8 and, since the numbers 2, 5, 8 each occur in only one of the sequences 2, 4; 3, 4, 7, 8; 5, 7; all prime implicants are required. Thus a simplest disjunctive normal form is

$$p \,\&\sim r \vee q \vee \sim p \,\&\, r$$

Let us, as a practical illustration of the technique for dealing with constraints, suppose that we wish to design a mechanism whose outputs correspond, in cyclic sequence, to the first four positive multiples of the binary number 11 (i.e. the decimal number 3). Thus, if outputs are fed back and the four digits of input and output correspond, in increasing order of significance, to p, q, r, s and to P, Q, R, S respectively, the relevant parts of the designation sequences are as follows:

p	T	F	T	F
q	T	T	F	F
r	F	T	F	T
s	F	F	T	T
P	F	T	F	T
Q	T	F	F	T
R	T	F	T	F
S	F	T	T	F

Thus we may write the designation sequences for P, Q, R, S as follows:

P	??0??1??0??1???
Q	??1??0??0??1???
R	??1??0??1??0???
S	??0??1??1??0???

and we may take $P = r$, $Q = r \equiv s$, $R = \sim r$, $S = r \not\equiv s$.

Examples 5D

1. Find a simple formula which corresponds to the designation sequence *FTTFFFTFT* if p, q, r cannot take the truth values F, T, F respectively. Find also simplest conjunctive and disjunctive normal forms of this formula, subject to the same constraint.

2. Design an adder for 3-bit binary numbers if it is known that the summands are prime, without leading zeros.
3. Design a multiplier (hardware) for two 3-bit binary numbers if it is known that the multiplier (number) is prime and the multiplicand is composite.

Solutions 5D

1. We must consider the derived value sequence

$$FT?FFTFT$$

If we regard ? as '*F*' this gives the formula

$$p \& (q \supset r)$$

For the conjunctive normal form we consider (by negation) the designation sequence

$$TF?TTFTF$$

with entries of *T* in the positions numbered 1, 4, 5, 7 and possibly also 3. Hence we then consider, as possible prime implicants, the formulae

$$\sim p \& \sim q \& \sim r \quad p \& q \& \sim r \quad \sim p \& \sim q \& r \quad \sim p \& q \& r$$

Thus we consider the following formulae, with verdicts as stated on the right:

$\sim p \& \sim q$	Consider further (1, 5)
$\sim p \& \sim r$	Consider further (1, 3)
$\sim q \& \sim r$	Reject (2)
$p \& q$	Reject (8)
$p \& \sim r$	Reject (2)
$q \& \sim r$	Consider further (3, 4)
$\sim p \& r$	Consider further (5, 7)
$\sim q \& r$	Reject (6)
$\sim p \& q$	Consider further (3, 7)
$q \& r$	Reject (8)

Hence none of the first four possible prime implicants is accepted (since the formulae $\sim p \& \sim q$, $q \& \sim r$, $\sim p \& q$ were marked 'consider further') and we now consider, for deletion of conjuncts, the formulae

$$\sim p \& \sim q \quad \sim p \& \sim r \quad q \& \sim r \quad \sim p \& r \quad \sim p \& q$$

We then reject the formulae q (8), $\sim q$ (2), r (6), $\sim r$ (2) and accept the formula $\sim p$ (1, 3, 5, 7). Therefore we must reject the formulae $\sim p \& \sim q$, $\sim p \& \sim r$, $\sim p \& r$, $\sim p \& q$ and the prime implicants are the formulae

$$\sim p \quad q \& \sim r$$

which deal with the positions numbered 1, 3, 5, 7; 3, 4 respectively. Both these prime implicants are essential (1, 4) so a simplest conjunctive normal form may be obtained by considering the formula

$$\sim (\sim p \lor q \& \sim r)$$

A simplest conjunctive normal form is thus the formula

$$p \& (\sim q \lor r)$$

2. The only possible summands are 5, 7 (101, 111 in binary notation). Thus the four columns below correspond to the four possible additions, the propositional variables p, q, r (s, u, v) corresponding to the digits of the first (second) summand and the formulae P, Q, R, S to the four digits of the sum, all in increasing order of significance.

$$
\begin{array}{cccc}
p & T & T & T & T \\
q & F & F & T & T \\
r & T & T & T & T \\
s & T & T & T & T \\
u & F & T & F & T \\
v & T & T & T & T \\
P & F & F & F & F \\
Q & T & F & F & T \\
R & F & T & T & T \\
S & T & T & T & T
\end{array}
$$

Since p, r, s, v always take the truth value T under the four relevant assignments, and q, u take successively under these assignments the respective truth values F, F; F, T; T, F; T, T, the designation sequences may be regarded as already written with respect to u and q (in that order). Thus we may take

$$ P = f \quad Q = q \equiv u \quad R = q \vee u \quad S = t $$

5.5 Application to the solution of simultaneous Boolean equations

If the Boolean operations $\cup$ and $\cap$ are rewritten $+$ and $.$ respectively we may, by means of the latter operations, write m equations in the n Boolean variables $X_1, ..., X_n$ where the coefficients are functions of other Boolean variables $A_1, ..., A_k$. Corresponding to the 2^k digits of the designation numbers of $A_1, ..., A_k$, we may write n designation numbers of 2^k digits for $X_1, ..., X_n$ as solutions of the equations. The number of admissible n-tuples of designation numbers will, of course, depend on the nature of the equations and it may be zero.

For example, the equations

$$ AX + B = Y $$

$$ AX' + B = Y' $$

in the variables X, Y have no pair of designation numbers as solutions since, if

$$ A, B = 1 $$

it follows from the given equations that

$$ Y = 1 = Y' $$

so that the numbers cannot have satisfactory last digits. On the other hand, the equations

$$(AX+B)\,Y = A+B$$

$$(A+B)\,XY = B+X$$

have as solutions

$$X = 0111 \quad Y = 0111 \qquad X = 0111 \quad Y = 1111$$

The general solution may, in all cases, be obtained by computing the designation numbers of both sides of the equations for each of the 2^n sets of values of $X_1, ..., X_n$.

For example, let us consider the equations

$$(AX+B)(CY'+A) = AB$$

$$AXY' + (B+C')\,Y = (A+C)'$$

and then a practical application. In each designation number we give the four possible values of each digit corresponding to the values (in that order) $0,0$; $0,1$; $1,0$; $1,1$ of X, Y respectively.

A		0000	1111	0000	1111	0000	1111	0000 1111
B		0000	0000	1111	1111	0000	0000	1111 1111
C		0000	0000	0000	0000	1111	1111	1111 1111
AX		0000	0011	0000	0011	0000	0011	0000 0011
$AX+B$		0000	0011	1111	1111	0000	0011	1111 1111
CY'		0000	0000	0000	0000	1010	1010	1010 1010
$CY'+A$		0000	1111	0000	1111	1010	1111	1010 1111
$(AX+B)(CY'+A)$		0000	0011	0000	1111	0000	0011	1010 1111
AB		0000	0000	0000	1111	0000	0000	0000 1111
AXY'		0000	0010	0000	0010	0000	0010	0000 0010
$B+C'$		1111	1111	1111	1111	0000	0000	1111 1111
$(B+C')\,Y$		0101	0101	0101	0101	0000	0000	0101 0101
$AXY'+(B+C')\,Y$		0101	0111	0101	0111	0000	0010	0101 0111
$(A+C)'$		1111	0000	1111	0000	0000	0000	0000 0000
Values for which digits match in the first equation	X	0011	00	0011	0011	0011	00	01	0011
	Y	0101	01	0101	0101	0101	01	11	0101
Values for which digits match in the second equation	X	01	0	01	0	0011	001	01	0
	Y	11	0	11	0	0101	011	00	0
Values for which digits match in both cases	X	01	0	01	0	0011	00	No	0
	Y	11	0	11	0	0101	01	cases	0

The equations are, in fact inconsistent. Although there are

$$2.1.2.1.4.2.1 \ (= 32)$$

ways (cf. 'values for which digits match in both cases' entries in the above table) of assigning pairs of values to the digits (other than the seventh ones) of the designation numbers of X, Y, we cannot make a satisfactory assignment for the seventh digits. The first equation gives

$$1010 \quad 0000$$

for

$$(AX+B)(CY'+A) \qquad AB$$

respectively, while the second gives

$$0101 \quad 0000$$

for

$$AXY'+(B+C')\,Y \qquad (A+C)'$$

respectively. Thus the first equation is satisfied only when $Y = 1$ and the second only when $Y = 0$.

On the other hand, the two equations, considered separately, have solutions. The first has $4, 2, 4, 4, 4, 2, 2, 4$ solutions for the digits numbered $1, ..., 8$ respectively, giving 8192 pairs of designation numbers as solutions, while the second has $2.1.2.1.4.3.2.1$ ($= 96$). An example of a solution to the first equation is (picking the first suitable pair of digits for X, Y in each case)

$$X = 00000000 \quad Y = 00000010$$

(We here write X as an abbreviation for 'the designation number of X' as there is no danger of ambiguity.)

Thus, if a mechanism has five inputs corresponding to the Boolean variables

$$A \quad B \quad C \quad X \quad Y$$

and an output corresponding to the Boolean expression

$$(AX+B)(CY'+A)$$

then this output will correspond also to

$$AB$$

irrespective of the values of A, B, C, provided that X, Y correspond to expressions with the above respective designation numbers. A solution is therefore provided by making X, Y correspond to the respective Boolean expressions

$$0 \quad A'BC$$

Expressions corresponding to any of the remaining 8191 pairs of designation numbers could, of course, have been used for X, Y, as for example

$$A'+B \quad 1$$

Examples 5E

1. Solve, if possible,
   (i) the equation $AXY' + B'(X + Y) = AB$
   (ii) the implication $AXY' + B'(X + Y) \supset AB$
   of Boolean algebra. (For (ii), if P, Q are formulae of the propositional calculus corresponding to the first and second Boolean expressions respectively, then $P \supset Q$ must take the truth value T. Thus the Boolean implication

   $$\Phi(A_1, ..., A_n) \supset \Psi(A_1, ..., A_n)$$

   is equivalent to the Boolean equation

   $$[\![\Phi(A_1, ..., A_n)]\!]' \cup \Psi(A_1, ..., A_n) = 1)$$

2. Solve, if possible, the simultaneous Boolean equations

   $$AXY' + (B + C')(X + YZ) = CA$$
   $$BX(CY + A) = B(A' + C)$$

3. A circuit has inputs corresponding to the Boolean variables A, B, X, Y and an output corresponding to the Boolean expression $AXY' + B'(X + Y)$. To what must the inputs X, Y correspond if the output always corresponds to AB? Give reasonably simple answers.

4. Repeat question 3 for outputs corresponding to (i) $AX + BY'$ and AB
   (ii) $AX + B(X' + Y)$ and AB

Solutions 5E

1. (ii)

A	0000	1111	0000	1111
B	0000	0000	1111	1111
AXY'	0000	0010	0000	0010
B'	1111	1111	0000	0000
$B'(X + Y)$	0111	0111	0000	0000
$AXY' + B'(X + Y)$	0111	0111	0000	0010
AB	0000	0000	0000	1111

Solutions
(rejecting $AXY' + B'(X + Y) = 1$ $\left.\begin{array}{l} X\ 0 \quad\ \ 0 \quad\ \ 0011 \quad 0011 \\ Y\ 0 \quad\ \ 0 \quad\ \ 0101 \quad 0101 \end{array}\right.$
$AB = 0$)

Thus there are 16 solutions, of which the first is

$$X = 0000 \quad Y = 0000$$

3. The four solutions to the equation of 1 (i) are

$$X = 00\alpha 1 \quad Y = 00\beta 0 \quad [\![\alpha, \beta \in \{0, 1\}]\!]$$

so simple solutions are (for $\alpha = 1, \beta = 0$)

$$X = B \quad Y = 0$$

5.6 Boolean matrices and application to change of variables

We found in Section 5.2 that we could always, given a formula, find the simplest conjunctive and disjunctive normal forms of that formula. We shall now extend the simplification process further to obtain what is known as the 'absolute simplest form'. This latter concept does not relate to a particular formula or designation sequence (or number) however, but to the class of designation sequences formed from a given designation sequence by rearrangement of its truth values. Thus, considering the Boolean representation of the formulae, for a 2^n-digit designation number with exactly u units it is sufficient to know the numerical values of n and u (rather than the value of the designation number) and there will be

$$(2^n)!/u!(2^n - u)!$$

designation numbers in the class. The problem is to determine, given the values of n and u, the simplest n-variable Boolean expression (formula of the propositional calculus) containing exactly u units (occurrences of T). We restrict algebraic operations to addition and multiplication, and logical operations correspondingly to disjunction and conjunction. If we do, in fact, wish to construct a simple formula having a given designation sequence, we need not conclude from the fact that the corresponding absolute simplest form has a different (i.e. rearranged) designation sequence, that its construction is of no help to us. The absolute simplest form may be adapted to the given designation sequence, after translation to its Boolean representative, by means of a change of variables. The theory of this change will be discussed, together with the theory of Boolean matrices, later in this section.

If a 2^n-digit designation number (corresponding to n Boolean or propositional variables) is the number of a Boolean expression containing N of these variables ($N \leq n$) then, provided that we make the irrelevant modification of regarding these as the first N variables, the given designation number will consist of a 2^N-element sequence of digits repeated 2^{n-N} times. Thus u will be an integral multiple of 2^{n-N} and, if

$$u = \sum_{i=k}^{n-1} 2^i x_{i+1} \quad (x_{k+1} = 1)$$

then

$$k \geq n - N$$

Thus

$$N \geq n - k$$

and the minimum number of binary operations in the Boolean expression is at least $n - k - 1$. This bound of

$$n - k - 1$$

is, in fact, always attained, as we shall now establish.

Let us consider the expression obtained from the expansion of u as

$$2^{n-1} x_n + \{2^{n-2} x_{n-1} + [\![2^{n-3} x_{n-2} + (\ldots + 2^k x_{k+1}) \ldots]\!]\}$$

and replace,* for decreasing values of i, the expressions

$$2^i x_{i+1} + ($$

by

$$A_{i+1} \cdot ($$

or by

$$A_{i+1} + ($$

according as

$$x_{i+1} = 0 \quad \text{or} \quad x_{i+1} = 1$$

Let us also replace

$$\ldots 2^k x_{k+1}) \ldots]\!]\} \quad \text{by} \quad A_{k+1}) \ldots]\!]\}$$

omitting the symbols beyond A_{k+1}. The formula corresponding to the resulting Boolean expression has a designation sequence containing $2^n - u$ F's and u T's, as we shall now prove by induction on n.

If $n = k+1$ then $u = 2^k$ and the formula is obtained from the expression

$$2^k . 1$$

Thus the formula is

$$p_{k+1} \quad \text{(corresponding to } A_{k+1})$$

and its designation sequence contains 2^{n-1} (i.e. 2^k or u) entries of T. We now assume the result for n and deduce it for $n+1$. If $x_{n+1} = 0$ the additional summand $2^n x_{n+1}$ does not affect the value of u and the new formula is

$$p_{n+1} \& P$$

where P is the formula for n. Since the formula $p_{n+1} \& P$ takes the truth value T if and only if the formulae p_{n+1}, P both take the truth value T, the formulae $p_{n+1} \& P, P$ will contain the same number of T's in their designation sequences. If $x_{n+1} = 1$ the additional summand increases the value of u by 2^n and the new formula is

$$p_{n+1} \vee P$$

whose designation sequence contains 2^n additional T's. Thus, by the induction hypothesis, the given formula contains exactly u occurrences of T.

---

* Other types of brackets may occur instead of ( ) and these should, of course, be similarly preserved when carrying out replacements.

As an illustration of the method, let us consider the construction of the absolute simplest form of the formula which contains 4 propositional variables and whose designation sequence contains 6 T's. Since

$$6 = \sum_{i=1}^{3} 2^i x_{i+1} \quad (x_2, x_3 = 1; \, x_4 = 0)$$

the formula is

$$p_4 \,\&\, (p_3 \vee p_2)$$

As a check we note that its designation sequence is

$$FFFFFFFFFFTTTTTT$$

If we are given a particular designation number for a Boolean expression corresponding to a formula it may be transformed to a designation number of an absolute simplest form by permuting the digits. This may, of course, be regarded as a permutation of the columns of the $1 \times n$ matrix (or row n-vector) formed† by the digits of the designation number. Thus, if 1 and 0 are regarded as integers, we can find a non-singular $n \times n$ matrix $\mathbf{T}$ such that, if $\mathbf{x}, \mathbf{y}$ denote the row vectors corresponding to the absolute simplest form's designation number and the given designation number respectively, then

$$\mathbf{y} = \mathbf{xT}$$

Since $\mathbf{T}$ is non-singular it follows at once that

$$\mathbf{x} = \mathbf{yT}^{-1}$$

Let us now suppose that a new Boolean expression is obtained from that corresponding to the absolute simplest form by replacing A_i by the expression A_i^* whose designation number (considered as a row vector) is obtained from that of A_i by post-multiplying it by $\mathbf{T}$ $(i = 1, ..., n)$. Since the effect of this multiplication is to rearrange the digits it follows at once that

$$A_i^* + A_j^* = (A_i + A_j)^* \quad (i, j = 1, ..., n)$$
$$A_i^* A_j^* = (A_i A_j)^* \quad (i, j = 1, ..., n)$$
$$(A_i^*)' = (A_i')^* \quad (i = 1, ..., n)$$

(Alternatively these three equations follow from the right distributive law for matrices.) Hence the new Boolean expression has the designation number corresponding to the row vector $\mathbf{y}$.

---

† Readers unfamiliar with the theory of matrices and elementary transformations should consult a suitable textbook, such as that by Birkhoff and McLane[25] before continuing with this section.

Let us, for example, consider the Boolean expression

$$A_1 A_2 + A_1'$$

whose designation number is

$$1011$$

The corresponding Boolean expression, with respect to the absolute simplest form of the related formula (i.e. the absolute simplest Boolean form) is

$$A_1 + A_2$$

with designation number

$$0111$$

Since

$$(1 \quad 0 \quad 1 \quad 1) = (0 \quad 1 \quad 1 \quad 1) \begin{pmatrix} 0 & 1 & 0 & 0 \\ 1 & 0 & 0 & 0 \\ 0 & 0 & 0 & 1 \\ 0 & 0 & 1 & 0 \end{pmatrix}$$

it follows that the designation number of $A_1{}^*$ is given by

$$(0 \quad 1 \quad 0 \quad 1) \begin{pmatrix} 0 & 1 & 0 & 0 \\ 1 & 0 & 0 & 0 \\ 0 & 0 & 0 & 1 \\ 0 & 0 & 1 & 0 \end{pmatrix} = (1 \quad 0 \quad 1 \quad 0)$$

and, similarly, that of $A_2{}^*$ is 0011. (The last two columns of $\mathbf{T}$ could have been taken as the last two columns of the unit matrix, since the last two digits of the designation number 1011 are the same. The interchange of the third and fourth columns is not pointless however since, had it not taken place, the designation number of $A_1{}^*$ would have been 1001 instead of 1010, leading ultimately to a more complicated answer which would not have been a genuine simplification of the original expression $A_1 A_2 + A_1'$.) Thus

$$A_1{}^* = A_1' \quad A_2{}^* = A_2$$

and

$$(A_1 + A_2)^* = A_1' + A_2$$

In the very simple cases where $n = 2$, simplifications such as that (of $A_1 A_2 + A_1'$ to $A_1' + A_2$) carried out above are often fairly obvious by inspection, but, in more complex cases, a less obvious simplification in circuitry is often effected (cf. Solutions 5F, 2(i)). Other applications of Boolean matrices are discussed by, for example, Ledley.[9]

Examples 5F

1. Find absolute simplest forms for the designation numbers for which

   (i) $\qquad n = 5 \quad u = 11 \quad (2^n = 32)$

   (ii) $\qquad n = 6 \quad u = 12$

2. Use the results of this section to simplify the circuits for the formulae

   (i) $\qquad p \,\&\, (q \equiv r) \vee (p/q)$

   (ii) $\qquad (p \equiv q \vee r) \supset (p \,\&\, r \vee q)$

3. Give the alternative proof of the three transformation equations using the distributive law.

Solutions 5F

1. (i) Since

$$11 = \sum_{i=0}^{4} 2^i x_{i+1} \quad (x_1, x_2, x_4 = 1; x_3, x_5 = 0)$$

the absolute simplest form is

$$A_5[\![A_4 + A_3(A_2 + A_1)]\!]$$

2. (i) The Boolean expression corresponding to the formula $p \,\&\, (q \equiv r) \vee (p/q)$ has the designation number

$$11101111$$

Thus $n = 3, u = 7$ and the absolute simplest form is

$$A_1 + A_2 + A_3$$

whose designation number is

$$01111111$$

Since $(1 \quad 1 \quad 1 \quad 0 \quad 1 \quad 1 \quad 1 \quad 1) =$

$$(0 \quad 1 \quad 1 \quad 1 \quad 1 \quad 1 \quad 1 \quad 1) \begin{pmatrix} 0 & 0 & 0 & 1 & 0 & 0 & 0 & 0 \\ 0 & 1 & 0 & 0 & 0 & 0 & 0 & 0 \\ 0 & 0 & 1 & 0 & 0 & 0 & 0 & 0 \\ 1 & 0 & 0 & 0 & 0 & 0 & 0 & 0 \\ 0 & 0 & 0 & 0 & 0 & 0 & 0 & 1 \\ 0 & 0 & 0 & 0 & 0 & 1 & 0 & 0 \\ 0 & 0 & 0 & 0 & 0 & 0 & 1 & 0 \\ 0 & 0 & 0 & 0 & 1 & 0 & 0 & 0 \end{pmatrix}$$

we may take $A_1{}^*$ to have the designation number

$$11001100$$

and thus we may take $A_1{}^*$ to be A_2'. Similarly we may take

$$A_2{}^* = A_1' \quad A_3{}^* = A_3$$

so that

$$(A_1 + A_2 + A_3)^* = A_2' + A_1' + A_3$$

Hence a decision mechanism for the Boolean expression $A_2' + A_1' + A_3$ or the formula $p \,\&\, q \supset r$ is sufficient.

3. Let the designation numbers of $A_i, A_i^*, A_i', (A_i^*)', (A_i')^*$ correspond to $\mathbf{x}, \mathbf{y}, \mathbf{z}, \mathbf{w}, \mathbf{t}$ respectively. Hence

$$= (1, ..., 1) - \mathbf{x}, \quad \mathbf{t} = \mathbf{z}\mathbf{T} = (1, ..., 1)\,\mathbf{T} - \mathbf{x}\mathbf{T} = (1, ..., 1) - \mathbf{x}\mathbf{T}$$

$$= (1, ..., 1) - \mathbf{y} = \mathbf{w}$$

Chapter 6

Problems Concerning Memory Elements

6.1 Further discussion of flip-flops—variations in function

In Chapter 4 we found that a flip-flop of a certain type could be constructed by a feedback technique, using a decision element for any one of four ternary functors. This particular type is often known as an R–S flip-flop or an S–R flip-flop (the letters R, S standing for 'reset' and 'set' respectively). It follows at once from our results that, if Q, R do not both take the truth value T, then

$$P \vee Q \nRightarrow R =_{\mathrm{T}} (P \nRightarrow R) \vee Q =_{\mathrm{T}} [\sim R, P, Q] =_{\mathrm{T}} L_2(P, Q, \sim R)$$

Thus, using Boolean notation, the variables A, A^*, B, C denoting the feedback, output, set and reset respectively, we may, using A, B, C as algebraic representatives of the formulae P, Q, R respectively and transforming the formula $(P \nRightarrow R) \vee Q$, write the conditions in the form

$$A^* = B + AC' \quad BC = 0$$

Two further forms of flip-flop are known as the D and T types which differ from the simple R–S version in one important respect. In addition to the normal input(s) they have a further one commonly known as the trigger, toggle or clock input which permits the flip-flop to change state only when it corresponds to the truth value T. In practice the signal is only momentarily in this state and is used to control the point in time when the flip-flop changes. Thus the normal input controls whether a change occurs and the toggle input determines when. Both D and T versions have only one normal input and a change of state occurs at the moment when the truth value corresponding to the toggle input changes from F to T provided that

the input and output correspond to opposite truth values—D type

the input corresponds to the truth value T—T type

The R–S flip-flop may, in certain circumstances, be used as a device known as the J–K flip-flop. It is now specified additionally that, if the inputs on the

141

set and reset wires both correspond to the truth value T then the output shall change from its previous state. Thus, if P, Q, R all take the truth value T then, with the notation of Section 4.8 $\Phi(P, Q, R)$ must take the truth value F and if P, Q, R take the truth values F, T, T respectively then $\Phi(P, Q, R)$ must take the truth value T. Hence, of the above four formulae, only

$$[\sim R, P, Q]$$

is now suitable.

6.2 Gating of memory elements and the associated simultaneous Boolean equations

We may write the equation

$$A^* = B + AC'$$

of Section 6.1 in the form

$$A_{n+1} = B_n + A_n C_n{}'$$

where A_n is the value of A after n units of time ($n = 0, 1, 2, ...$), similar notations being used for other variables. The discussion of flip-flops in Chapter 4 does not involve the concept of a unit of time but, when using hardware in which a chain of successive pulses provide a time reference, we can always arrange for a fed-back output to be regarded as the value of that output one unit of time earlier, though a detailed discussion of the methods employed is beyond the scope of this book (cf. Section 1.9).

Memory devices may be combined with ordinary decision elements to perform certain functions and we may, by using the above notation, construct simultaneous Boolean equations whose algebraic solution corresponds to the decision mechanisms which we must build. These equations may then be solved by the methods of the previous chapter. Let us, for example, suppose that we wish to design a counter whose outputs correspond, in succession, to the numbers

$$0, 3, 1, 2, 0, ...$$

each number occurring in every fourth place of the sequence, and that the two digits of these numbers (considered as the binary numbers 00, 11, 01, 10, 00, ...) are the outputs of two available R–S flip-flops. The problem is then to build decision mechanisms using these two flip-flops and ordinary decision elements.

Let us associate with the flip-flop whose outputs correspond to the less significant digits of the numbers of the above sequence the variables

$$A_n, B_n, C_n \quad (n = 0, 1, ...)$$

as above and, with the other flip-flop, the variables

$$D_n, E_n, F_n \quad (n = 0, 1, \ldots)$$

the roles of A_n, B_n, C_n being taken over by D_n, E_n, F_n respectively $(n = 0, 1, \ldots)$. We must therefore arrange for the values of A_n to correspond to the elements of the sequence

$$0, 1, 1, 0, 0, \ldots$$

the value of A_n corresponding to the $(i+1)$th element $(i = 0, 1, \ldots)$, and the values of D_n to correspond, in a similar manner, to the elements of the sequence

$$0, 1, 0, 1, 0, \ldots$$

We note first that

$$A_{n+1} = 1$$

if and only if

$$A_n = D_n$$

and that

$$D_{n+1} = 1$$

if and only if

$$D_n = 0$$

$(n = 0, 1, \ldots)$. Thus we may write

$$\left. \begin{array}{l} A_{n+1} = A_n D_n + A_n{}' D_n{}' \\ D_{n+1} = D_n{}' \end{array} \right\} \quad (n = 0, 1, \ldots)$$

However, we found at the beginning of this section that

$$A_{n+1} = B_n + A_n C_n{}' \quad (n = 0, 1, \ldots)$$

and it follows similarly that

$$D_{n+1} = E_n + D_n F_n{}' \quad (n = 0, 1, \ldots)$$

Hence, if we can express B_n, C_n, E_n, F_n in terms of A_n, B_n in such a way that

$$\left. \begin{array}{r} A_n D_n + A_n{}' D_n{}' = B_n + A_n C_n{}' \\ D_n{}' = E_n + D_n F_n{}' \end{array} \right\} \quad (n = 0, 1, \ldots) \qquad \begin{array}{l} (1) \\ (2) \end{array}$$

we have only to feed outputs from the corresponding decision mechanisms for B_n, C_n, E_n, F_n to the set and reset inputs of the flip-flops, provided, of course, that our solutions satisfy the R–S flip-flop conditions

$$\left. \begin{array}{l} B_n C_n = 0 \\ E_n F_n = 0 \end{array} \right\} \quad (n = 0, 1, \ldots) \qquad \begin{array}{l} (3) \\ (4) \end{array}$$

Let us consider first the equations (1), (3) as simultaneous Boolean equations in the variables B_n, C_n. Applying the methods of Chapter 5, we find that the general solution (with respect to D_n and A_n in that order) is $[\![x, y \in \{0, 1\}]\!]$

$$B_n = 100y$$

$$C_n = 0x10$$

and, taking

$$x, y = 0$$

we obtain the solution

$$B_n = A_n{}'D_n{}' \quad C_n = A_n D_n{}'$$

Similarly equations (2) and (4) give

$$E_n = 1010$$

$$F_n = 0101$$

as the only solution. Thus

$$E_n = D_n{}' \quad F_n = D_n$$

Substituting back in our flip-flop equations it follows at once that

$$A_{n+1} = A_n{}'D_n{}' + A_n(A_n{}' + D_n)$$

$$D_{n+1} = D_n{}' + D_n D_n{}'$$

so it is sufficient to connect the set and reset inputs of the flip-flops to decision mechanisms for $A_n{}'D_n{}', A_n D_n{}', D_n{}', D_n$ as shown in Figure 6.1.

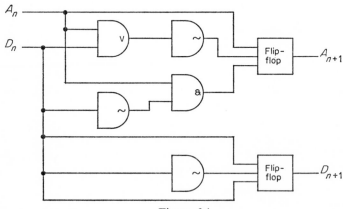

Figure 6.1

It is important to remember that the equations (3), (4) must be satisfied, so we cannot simplify the last two equations to

$$A_{n+1} = A_n' D_n' + A_n D_n$$
$$D_{n+1} = D_n' + D_n \cdot 1'$$

corresponding to the 'solution'

$$B_n = A_n' D_n' \quad C_n = D_n' \quad E_n = D_n' \quad F_n = 1$$

since, then

$$B_n C_n = A_n' D_n' \quad E_n F_n = D_n'$$

and, for $n = 0$,

$$A_n' D_n' = D_n' = 1 \neq 0$$

If $B_n C_n \neq 0$ or $E_n F_n \neq 0$ we have no guarantee that the flip-flop difference equations are valid.

Examples 6A

1. Obtain Boolean difference equations in (some or all of) the variables A_n, A_{n+1}, B_n, C_n for D, T and J–K flip-flops.
2. Redesign the counter of Section 6.2, using two T flip-flops.
3. Design, if possible, a 3 flip-flop counter for the sequence

$$0, 5, 3, 2, 7, 1, 6, 4, 0, \ldots$$

of 3-digit binary numbers, possibly with leading zeros, using (i) R–S flip-flops, (ii) J–K flip-flops, (iii) T flip-flops.

Solutions 6A

1. For T flip-flops, using B_n, C_n, corresponding to the input and toggle respectively

$$A_{n+1} = (A_n' B_n + A_n B_n') C_n + A_n C_n'$$

For D flip-flops, using B_n, C_n as immediately above,

$$A_{n+1} = B_n C_n + A_n C_n'$$

2.
$$A_{n+1} = A_n D_n + A_n' D_n',$$

(from text, using D_n, E_n, F_n in place of A_n, B_n, C_n respectively for the second flip-flop)

$$A_{n+1} = A_n' B_n + A_n B_n'$$

(from previous solution, since we may take $C_n = 1$).
Thus we must find B_n such that

$$A_n' B_n + A_n B_n' = A_n D_n + A_n' D_n'$$

This equation has the unique solution

$$B_n = 1010$$

Thus
$$B_n = D_n'$$
Similarly
$$D_{n+1} = D_n'.$$
$$D_{n+1} = D_n'E_n + D_n E_n'$$
and the equation
$$D_n' = D_n'E_n + D_n E_n'$$
has the unique solution
$$E_n = 1111$$
Thus
$$E_n = 1$$
and the two inputs correspond to D_n' and 1.

6.3 Sequential circuit analysis

In Section 6.2 we considered the design of circuits involving both decision elements and flip-flops, in the case where the number of flip-flops was given *a priori*. It is often advantageous to keep the number of flip-flops to a minimum to reduce problems of time delays, and it is therefore instructive to consider the redesign of a circuit so as to contain as few flip-flops as possible, subject, of course, to the requirement that the circuit can still carry out its original function. Some methods, due to Huffman[26] and Mealy,[27] for reducing the number of flip-flops will now be considered.

Let us, for example, consider a circuit with an input wire from another mechanism and an output wire whose states correspond to the values of the Boolean variables A_n, B_n respectively ($n = 0, 1, ...$) and two flip-flops whose outputs correspond to the variables C_n, D_n. Suppose also that the values of B_n, C_{n+1}, D_{n+1} are determined by the values of A_n, C_n, D_n ($n = 0, 1, ...$). We shall consider here the particular example in which these three determinations are as shown below, the case where
$$C_n = 1 \quad D_n = 0$$
arising for no non-negative integer n.

A_n	C_n	D_n	C_{n+1}	D_{n+1}	B_n
0	0	0	0	1	0
0	1	0	Does not arise		
0	0	1	0	1	0
0	1	1	0	0	1
1	0	0	0	1	0
1	1	0	Does not arise		
1	0	1	0	0	0
1	1	1	1	1	0

We note now that the values of C_n and D_n relate only to the internal states of the flip-flops and not to inputs to, or outputs from, the whole mechanism. Thus we may rename the three pairs of states (the fourth pair $C_n = 1, D_n = 0$ being redundant)

a (previously $C_n = 0, D_n = 0$)

b (previously $C_n = 0, D_n = 1$)

c (previously $C_n = 1, D_n = 1$)

and draw up the table accordingly.

New state and B_n	a	b	c	State
0	$b, 0$	$b, 0$	$a, 1$	
1	$b, 0$	$a, 0$	$c, 0$	
A_n				

We then note that if, in the above table, states a and b are identified, we obtain the table shown below, in which the first two columns are identical.

New state and B_n	a	a	c	State
0	$a, 0$	$a, 0$	$a, 1$	
1	$a, 0$	$a, 0$	$c, 0$	
A_n				

Thus we may identify states a and b, without contradiction, and the above table (after deletion of its redundant second column) becomes that shown below, using the original notation. D_n is not needed when there are only two states. (We rename the states a, c as the states where $C_n = 0, 1$ respectively, though the opposite renaming would have been equally suitable.)

A_n	C_n	C_{n+1}	B_n
0	0	0	0
0	1	0	1
1	0	0	0
1	1	1	0

If, on the other hand, the third line of the original table had given the value of C_{n+1} as 1 instead of 0, the new table would have been as shown below, so

New state and B_n	a	b	c	State
0	$b, 0$	$c, 0$	$a, 1$	
1	$b, 0$	$a, 0$	$c, 0$	
A_n				

that the first two entries in the first row would have become contradictory on identifying states a and b. Similarly we should have obtained contradictions in the first row on identifying states a, c and also on identifying states b, c. The only other possibility would be the identification of all three states, in which case, as shown below, the last two entries in the first row would have

New state and B_n	a	a	a	State
0	$a, 0$	$a, 0$	$a, 1$	
1	$a, 0$	$a, 0$	$a, 0$	
A_n				

become contradictory. Thus all three states a, b, c are essential and we cannot make do with less than three flip-flops.

As the number of flip-flops initially under consideration increases, the number of proper subsets of the set of flip-flops (and therefore the number of internal states $a, b, c, \ldots$) increases rapidly. Two rules for identifying states have been given by Huffman and Mealy.[26, 27] Although these rules do not always reduce the number of states to the minimum possible, they are very useful and can be guaranteed to give a reduction to the minimum number possible when there are no redundancies, other than those of the type considered in this section. A good account of these rules is given by Phister.[23]

Examples 6B

1. Simplify, if possible, by reducing the number of flip-flops to the minimum, the circuit corresponding to the table given below.

A_n	C_n	D_n	C_{n+1}	D_{n+1}	B_n
0	0	0	0	0	0
0	1	0	1	1	1
0	0	1	1	0	1
0	1	1	1	1	1
1	0	0	Does not arise		
1	1	0	0	1	0
1	0	1	1	0	0
1	1	1	1	0	0

2. Simplify, if possible, by reducing the number of flip-flops to the minimum, the circuit corresponding to the table given below (i) if $x = 0$, (ii) if $x = 1$.

A_n	C_n	D_n	C_{n+1}	D_{n+1}	B_n
0	0	0	0	0	0
0	1	0	1	1	1
0	0	1	1	0	1
0	1	1	1	1	1
1	0	0	Does not arise		
1	1	0	0	1	0
1	0	1	1	0	0
1	1	1	1	0	x

Solutions 6B

1. If we rename the pairs of states of the two flip-flops

a (previously $C_n = 0$, $D_n = 0$)
b (previously $C_n = 1$, $D_n = 0$)
c (previously $C_n = 0$, $D_n = 1$)
d (previously $C_n = 1$, $D_n = 1$)

we obtain the table given below.

New state and B_n	a	b	c	d	State
0	$a, 0$	$d, 1$	$b, 1$	$d, 1$	
1	—	$c, 0$	$b, 0$	$b, 0$	
A_n					

Identifying the states b, c, d we obtain the following table. This is the simplest possible, since the identification of state a with one or more of the states b, c, d would cause contradictory entries in the first row of the table.

New state and B_n	a	b	State
0	$a, 0$	$b, 1$	
1	—	$b, 0$	
A_n			

If we then interpret the new states a, b as $C_n = 0, 1$ respectively, we obtain the table shown below.

A_n	C_n	C_{n+1}	B_n
0	0	0	0
0	1	1	1
1	0	—	—
1	1	1	0

Any triple of entries could have been made for the blanks in the original table, but this would not help matters in the present example (see above).

6

6.4 Sequential circuit synthesis

If we wish to design, in the best possible way, a circuit with certain properties, we may begin by constructing a particular solution to the problem, assigning internal states as situations occur and remembering that, if a situation is repeated, a new internal state is not needed. The resulting solution may then be simplified by the methods of Section 6.3.

Let us suppose, for example, that a mechanism has an input wire which receives three input signals A_0, A_1, A_2 corresponding to the digits, in increasing order of significance, of an integer x such that $0 \leqslant x \leqslant 7$, x being regarded as a binary number. We shall suppose further that there is an output wire whose signal corresponds to 0 at least until the final digit of x is signalled and that, at this point, the output signal corresponds to 1 if and only if x has one of the values 5, 7 (i.e. 101, 111 in binary notation). The circuit is to repeat the test for 5 and 7 indefinitely on further (serial) input triples.

The initial internal state will be denoted by a and, with the notation of Section 6.3, the next internal state will be denoted by b or c according as A_0 is equal to 0 or to 1. In the former case the following internal state is denoted by d or e according as A_1 is equal to 0 or to 1 and f, g are defined similarly with respect to the latter case. Thus, after three input signals have been received, the output will correspond to 1 if and only if the first and third signals corresponded to 1 (irrespective of the value of the second input signal), i.e. if and only if the current internal state is f or g and $A_2 = 1$. After three input signals have been received the test is complete and the internal state is one of the states d, e, f, g. Thus the internal state a must follow all the internal states d, e, f, g irrespective of the value of A_2, so that the next triple of input signals can be tested. The initial table is therefore as shown below.

Next state and B_n	a	b	c	d	e	f	g	State
0	$b, 0$	$d, 0$	$f, 0$	$a, 0$	$a, 0$	$a, 0$	$a, 0$	
1	$c, 0$	$e, 0$	$g, 0$	$a, 0$	$a, 0$	$a, 1$	$a, 1$	
A_n								

We observe next that, if we identify states d, e and also states f, g the table becomes that shown below, in which there are no contradictions.

Next state and B_n	a	b	c	d	d	f	f	State
0	$b, 0$	$d, 0$	$f, 0$	$a, 0$	$a, 0$	$a, 0$	$a, 0$	
1	$c, 0$	$d, 0$	$f, 0$	$a, 0$	$a, 0$	$a, 1$	$a, 1$	
A_n								

The state f cannot be identified with any of the states a, b, c, d in view of the nature of the last entry on the lower row of the new table. Since state c always changes to f after one unit of time, the former state cannot be identified with any of the states a, b, d and similar arguments show that no further simplification is possible.

Examples 6C

1. Design a mechanism similar to that described in the text of Section 6.4 in which the roles of 5, 7 are taken over by 3, 4 and 6, simplifying the result as much as possible.
2. Design a mechanism similar to that described in the text of Section 6.4 for which $B_n = 1$ if and only if

$$A_n = A_{n-1} = A_{n-2} \quad (n = 2, 3, \ldots)$$

and

$$B_0 = B_1 = 0$$

simplifying the result as much as possible.
3. Repeat question 2 for the case where $B_n = 1$ if and only if the binary number formed by A_n, A_{n-1} (A_n corresponding to the more significant digit) is less than that correspondingly formed by A_{n-2}, A_{n-3} ($n = 3, 4, \ldots$) and

$$B_0 = B_1 = B_2 = 0$$

Solutions 6C

1. Our initial table is as shown below. Clearly, states d, e may be identified, but, by methods similar to those used in Section 6.4, no further simplification is possible.

New state and B_n	a	b	c	d	e	f	g	State
0	$b, 0$	$d, 0$	$f, 0$	$a, 0$	$a, 0$	$a, 0$	$a, 1$	
1	$c, 0$	$e, 0$	$g, 0$	$a, 1$	$a, 1$	$a, 0$	$a, 0$	
A_n								

2. Let the initial internal state be a and let the state $b(c)$ correspond to the case where a single digit (but not more than this) of 0(1) has been found immediately previously. Let the state $d(e)$ correspond to the case where two consecutive digits of 0(1) have been found immediately previously and the states f, g correspond similarly to triples of 0's and 1's. (Since we are never concerned with happenings more than three digits in arrear the states f, g correspond to runs of four or more equal digits also.) Our initial table is therefore as shown below.

New state and B_n	a	b	c	d	e	f	g	State
0	$b, 0$	$d, 0$	$b, 0$	$f, 1$	$b, 0$	$f, 1$	$b, 0$	
1	$c, 0$	$c, 0$	$e, 0$	$c, 0$	$g, 1$	$c, 0$	$g, 1$	
A_n								

Identifying states d, f with e, g respectively we obtain, without contradiction, the table shown below.

New state and B_n	a	b	c	d	e	State
0	$b, 0$	$d, 0$	$b, 0$	$d, 1$	$b, 0$	
1	$c, 0$	$c, 0$	$e, 0$	$c, 0$	$e, 1$	
A_n						

Methods similar to those of Section 6.4 show that no further simplification is possible.

6.5 The state assignment problem

At the end of Section 6.3 we gave a solution to our problem in which the states a, c were renamed as $C_n = 0, 1$ respectively. If we had renamed a, c as $C_n = 1, 0$ respectively we would have obtained the table given below.

A_n	C_n	C_{n+1}	B_n
0	0	1	1
0	1	1	0
1	0	0	0
1	1	1	0

The difference equations in the two cases are of course not the same, and different logic and circuits will in general result.

Some circuits will be more complex than others, though the simplest may not necessarily be the best. If two flip-flops should be set simultaneously and there is, in fact, a slight lack of synchronization, certain intermediate (unstable) internal states may occur when these are not intended. The stable states ultimately reached may, as a consequence of this, be incorrect, though this is not always the case. However, hazards of this type must be borne in mind when interpreting the logic, to ensure that a reliable device (for a control mechanism or other purpose) is obtained. A fuller discussion of these problems is given by Caldwell.[28]

Examples 6D

1. Design circuits corresponding to the two possible state assignments in the solution to Examples 6B, 1.
2. Solve the corresponding problems for Examples 6B, 2(i) and (ii).

Solutions 6D

1. If a is $C_n = 0$ we consider the last table of Solutions 6B, 1. Thus $C_{n+1} = C_n$, $B_n = A_n{'}C_n$.

Chapter 7

Automata and Methods for Their Construction

7.1 The concept of an automaton

We have now considered a number of uses of decision elements for the construction of mechanisms, both in cases where the state of the output depends only on the present (or immediate past) states of the inputs and in cases where it depends also on earlier states of inputs. For example, a decision mechanism for the formula $p \& q \lor \sim (r \supset p)$ is a mechanism of the former type and a flip-flop is a mechanism of the latter type. We are thus led to the consideration of the more general concept of an automaton.

A (deterministic) finite automaton is a device with a finite number of inputs and a finite number of outputs, each of which is capable of two physical states (which we may, as in Chapter 1, regard as corresponding to the truth values T, F). The automaton is itself capable of assuming, at a given time, any one of a finite number of physical states and the state of each output is determined solely by the states of the inputs and the internal state of the automaton. Strictly speaking (cf. Chapter 6) the output may be in the corresponding state after a lapse of one unit of time. We have already met in Chapter 6 the concept of an internal state, but we then restricted ourselves to combining flip-flops and ordinary decision elements. In this chapter we consider sequential machinery in a more general way, the flip-flops considered in previous chapters being particular illustrations of the present more general theory. An automaton may also have an infinite external memory. Such an infinite automaton is known as a *Turing machine*. Such machines[29] are of great interest in connexion with the theory of effective calculability and unsolvable problems, but a study of this branch of Mathematical Logic is beyond the scope of this book. We shall, in future, refer implicitly to *finite* automata when we speak simply of automata.

Those automata with only one output whose state depends only on the states of the inputs (and not on the internal state of the automaton) are, of course, decision mechanisms (cf. Chapter 1). Naturally, if we remove the restriction that there shall be only one output, we obtain merely a collection of decision mechanisms. Devices such as flip-flops and adders are particular cases of the more general concept of an automaton, since the dependence of the state of an output on previous states of inputs may, as we shall see, be brought about by considering the internal state of the automaton.

7.2 Nerve nets

A particular type of example of a finite automaton is a McCulloch–Pitts net of *nerve cells* or *neurons*. A number of important results concerning the applicability of these nets to particular situations have been published by Kleene[12] in a partly expository article. Most of these are beyond the scope of this book, but we shall now derive a few of the simpler results as consequences of results of Chapters 1 and 2.

A neuron is effectively a decision element whose output assumes a state corresponding to the states of the inputs one unit of time after those inputs have assumed their states. It corresponds to a functor $F_{abc}(,...,)$ of $a+b$ arguments ($a = 1, 2, ...,; b = 0, 1, ...; c = 1, ..., a$) whose truth table is determined as follows. Let $L_{ij}(,...,)$ $[\![M_{kj}(,...,)]\!]$ be a functor of j arguments ($i = 1, ..., j; k = 0, ..., j-1; j = 1, 2, ...$) such that the formula $L_{ij}(P_1, ..., P_j)$ $[\![M_{kj}(P_1, ..., P_j)]\!]$ takes the truth value T if and only if at least (most) $i(k)$ of the formulae $P_1, ..., P_j$ take the truth value T. Then

$$F_{abc}(P_1, ..., P_{a+b}) =_T L_c(P_1, ..., P_a) \,\&\, M_0(P_{a+1}, ..., P_{a+b})$$

Thus the formula $F_{abc}(P_1, ..., P_{a+b})$ takes the truth value T if and only if the formulae $P_{a+1}, ..., P_{a+b}$ all take the truth value F and at least c of the formulae $P_1, ..., P_a$ take the truth value T. The number c is known as the *threshold* of the neuron. We note that conjunction, disjunction and non-implication are particular examples of neuron-corresponding functors since

$$\text{(a)} \quad F_{202}(P, Q) =_T P \,\&\, Q$$
$$\text{(b)} \quad F_{201}(P, Q) =_T P \vee Q$$
$$\text{(c)} \quad F_{111}(P, Q) =_T P \not\supset Q$$

We note that, more generally,*

$$\prod_{i=1}^{\alpha} P_i \,\&\, \prod_{i=\alpha+1}^{\alpha+\beta} \sim P_i =_T F_{\alpha\beta\alpha}(P_1, ..., P_{a+\beta}) \quad (\alpha = 1, 2, ...; \beta = 0, 1, ...) \quad \text{(A)}$$

$$\sum_{i=1}^{\alpha} P_i =_T F_{\alpha 01}(P_1, ..., P_\alpha) \quad (\alpha = 1, 2, ...) \quad \text{(B)}$$

---

* If $\beta = 0$ the conjunction functor and second conjunct of (A) are, of course, omitted.

We are now in a position to show how to construct, given positive integers l, n, an automaton with n inputs and one output, such that the state of the output at a given time is determined, in a prescribed way, by the states of the inputs $2, ..., l+1$ units of time previously. Let the state of the ith input k units of time previously correspond to the truth value of the propositional variable denoted by the syntactical variable P_{ik} ($i = 1, ..., n$; $k = 2, ..., l+1$). The state of the output must then correspond to the truth value of a formula

$$\Phi(P_{12}, ..., P_{1,l+1}, ..., P_{n2}, ..., P_{n,l+1})$$

The latter formula must have a disjunctive normal form, which we shall denote by Q.

We shall, for the present, assume that Q takes the truth value F when all the nl propositional variables occurring in it take the truth value F. Thus every disjunct of Q must contain at least one unnegated propositional variable and it follows at once from (A) (which is valid for all positive integers α) that every disjunct may be represented by a neuron, provided that all input signals corresponding to the truth values of

$$P_{12}, ..., P_{1,l+1}, ..., P_{n2}, ..., P_{n,l+1}$$

are available simultaneously. Thus (except in one trivial case, for which a solution is obvious), by (B), Q may be represented by using the outputs of the disjunct-corresponding neurons as inputs to a further neuron. The output is thus obtained two units of time after the inputs corresponding to

$$P_{12}, ..., P_{1,l+1}, ..., P_{n2}, ..., P_{n,l+1}$$

are fed into the automaton.

The problem is therefore solved in the case $l = 1$ but, in general, inputs corresponding to $P_{13}, ..., P_{1,l+1}, ..., P_{n3}, ..., P_{n,l+1}$ will be obtained too soon and must be preserved. An output may be preserved for λ units of time by feeding it through λ delay mechanisms of one unit in series, since such a mechanism may be obtained from a neuron in view of the validity of the equation

$$P =_{\text{T}} F_{101}(P) \tag{C}$$

which is a special case of (A). In this case the inputs corresponding to $P_{13}, ..., P_{1,l+1}, ..., P_{n3}, ..., P_{n,l+1}$ are available as the outputs of delay mechanisms and the states of these mechanisms together constitute the 'internal state' of the automaton.

If Q takes the truth value T when all the nl propositional variables take the truth value F then, under this assignment, $\sim Q$ takes the truth value F as does any disjunctive normal form of $\sim Q$. Thus we may use the above method again provided that we invert the interpretations of the two possible output signals.

Let us, for example, suppose that we wish to construct an automaton with two inputs and one output. The output after p units of time ($p = 2, 3, ...$) or, as we shall now say, at time p, indicating that, from times $p-4$ to $p-2$ (both inclusive)

(i) the state of the second input has corresponded to the truth value T whenever the state of the first input has corresponded to the truth value T

and

(ii) the state of the second input has corresponded to the truth value F at least once.

Clearly in this case

$$n = 2 \quad l = 3$$

and we must construct a formula $\Phi(P_{12}, P_{13}, P_{14}, P_{22}, P_{23}, P_{24})$. The first condition requires that the formulae

$$P_{1i} \supset P_{2i} \quad (i = 2, 3, 4)$$

take the truth value T and the second that the formula

$$\sim \prod_{i=2}^{4} P_{2i}$$

takes the truth value T. Thus, omitting associations, a suitable choice for the formula $\Phi(P_{12}, ..., P_{24})$ is

$$(P_{12} \supset P_{22}) \,\&\, (P_{13} \supset P_{23}) \,\&\, (P_{14} \supset P_{24}) \,\&\, \sim (P_{22} \,\&\, P_{23} \,\&\, P_{24})$$

As this formula takes the truth value T when all six propositional variables take the truth value F, we must reduce its negation to disjunctive normal form. We note that, since $\sim (P \supset Q) =_T P \,\&\, \sim Q$, it follows from the de Morgan laws that

$$\sim \Phi(P_{12}, ..., P_{24}) =_T P_{12} \,\&\, \sim P_{22} \vee P_{13} \,\&\, \sim P_{23} \vee P_{14} \,\&\, \sim P_{24} \vee P_{22} \,\&\, P_{23} \,\&\, P_{24}$$

The required mechanism is then as shown in Figure 7.1, the inputs corresponding to $P_{13}, P_{14}, P_{23}, P_{24}$ being obtained as the outputs of delay mechanisms, using (C). The particular applications of (A) and (B) in this case are

$$P_{1i} \,\&\, \sim P_{2i} =_T F_{111}(P_{1i}, P_{2i}) \quad (i = 2, 3, 4)$$

$$P_{22} \,\&\, P_{23} \,\&\, P_{24} =_T F_{303}(P_{22}, P_{23}, P_{24})$$

$$P_{12} \,\&\, \sim P_{22} \vee P_{13} \,\&\, \sim P_{23} \vee P_{14} \,\&\, \sim P_{24} \vee P_{22} \,\&\, P_{23} \,\&\, P_{24}$$

$$=_T F_{401}(P_{12} \,\&\, \sim P_{22}, P_{13} \,\&\, \sim P_{23}, P_{14} \,\&\, \sim P_{24}, P_{22} \,\&\, P_{23} \,\&\, P_{24})$$

The output represents the conditions in the sense that its state corresponds to the truth value F if and only if the states of the inputs correspond to the conditions being met.

If we weaken somewhat the requirement that the state of the output at time p be determined by the states of the inputs at times $p-l-1,...,p-2$ we may impose important restrictions on the values of the suffixes in the

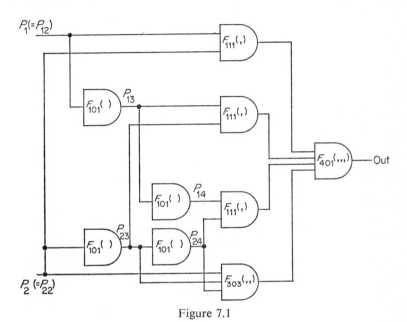

Figure 7.1

above functors $F_{abc}(,...,)$. Let us now suppose that for some (constant) integer D, the state of the output is determined, in a prescribed way, by the states of the inputs at times $p-l-D-1,...,p-D-2$. We shall show that, given this prescribed way, we may always for some appropriate value of D, construct an automaton using only neurons corresponding to functors $F_{abc}(,...,)$ such that

$$a+b \leqslant 2 \quad \text{(and therefore } c \leqslant 2)$$

Thus all neurons correspond to functors of not more than two arguments.

It will be convenient to consider the state of the output at time $p+D$ to be determined by the states of the inputs at times

$$p-l-1,...,p-2 \quad (p = l+1, l+2, ...)$$

Thus we may again* regard the truth value of P_{ik} as corresponding to the state of the ith input at time $p-k$ ($i = 1, ..., n$; $k = 2, ..., l+1$). We note that the disjunctive normal form Q considered in the previous case is of the form

$$\sum_{i=1}^{\tau} S_i$$

and that each of the disjuncts S_i of the set $\{S_1, ..., S_\tau\}$ is, for some integers α, β ($\alpha = \alpha(i)$, $\beta = \beta(i)$, $1 \leqslant \alpha \leqslant \beta$) of the form

$$\prod_{j=1}^{\alpha} R_j \,\&\, \prod_{j=\alpha+1}^{\beta} \sim R_j$$

Since

(i) $\displaystyle\prod_{i=1}^{\gamma} R_i =_{\mathrm{T}} \prod_{i=1}^{\gamma-1} R_i \,\&\, R_\gamma$ ($\gamma = 2, 3, ...$)

(ii) $\displaystyle\prod_{i=1}^{\alpha} R_i \,\&\, \prod_{i=\alpha+1}^{\gamma} \sim R_i =_{\mathrm{T}} \left(\prod_{i=1}^{\alpha} R_i \,\&\, \prod_{i=\alpha+1}^{\gamma-1} \sim R_i \right) \dotplus R_\gamma$ ($\gamma = \alpha+1, \alpha+2, ...$)

(iii) $\displaystyle\sum_{i=1}^{\gamma} S_i =_{\mathrm{T}} \sum_{i=1}^{\gamma-1} S_i \vee S_\gamma$ ($\gamma = 2, 3, ...$)

it follows at once from (i), (ii), (iii) that, if time could be ignored, we could construct the required decision mechanism using only neurons with not more than two inputs.

Let us suppose that the decision mechanism is constructed in such a way that no output is connected to more than one input† and that the input corresponding to P_{ik} is connected to the final output through exactly λ_{ik} neurons in series ($i = 1, ..., n$; $k = 2, ..., l+1$). We shall show how to add delay units to the mechanism of the previous paragraph, so as to eliminate troubles due to the time factor. It may happen, as we shall see in the example discussed below, that if an input is connected to more than one neuron, the value of λ_{ik} is not determined uniquely by the values of i and k, but in this case we may connect the input corresponding to P_{ik} to more than one series of delay mechanisms.

Since there are λ_{ik} neurons in series between the input corresponding to P_{ik} and the final output, we require this output to be available at time $p + D - \lambda_{ik}$ ($i = 1, ..., n$; $k = 2, ..., l+1$). In fact this input becomes available

---

* At times $0, ..., l+D$ the output will be determined similarly but P_{ik} must be regarded as taking the truth value F when $k < 0$.

† This can always be accomplished by duplicating decision mechanisms for sub-formulae with more than one occurrence in the given formula.

at time $p-k$ and can be made available at time $p-k+\mu$ by using μ delay units in series ($\mu = 1, 2, ...$), but it cannot be made available at time $p-k-\nu$ ($\nu = 1, 2, ...$). Thus we must assign to D a value such that

$$p + D - \lambda_{ik} \geqslant p - k$$

that is,

$$D \geqslant \lambda_{ik} - k \quad (i = 1, ..., n; \ k = 2, ..., l+1)$$

Thus, if the values of λ_{ik} are $\phi_{ik1}, ..., \phi_{ikm_{ik}}$ and

$$\theta_{ik} = \max (\phi_{ik1}, ..., \phi_{ikm_{ik}})$$

we may assign to D the value

$$\max (\theta_{12} - 2, ..., \theta_{1,l+1} - l - 1, ..., \theta_{n2} - 2, ..., \theta_{n,l+1} - l - 1)$$

and delay, using (C), the signal corresponding to P_{ik} by $D - \lambda_{ik} + k$ units of time for each of the m_{ik} values of λ_{ik} ($i = 1, ..., n; \ k = 2, ..., l+1$).

In the particular example considered previously we obtained the disjunctive normal form

$$P_{12} \ \& \sim P_{22} \vee P_{13} \ \& \sim P_{23} \vee P_{14} \ \& \sim P_{24} \vee P_{22} \ \& \ P_{23} \ \& \ P_{24}$$

which may, by (i), (ii), (iii), (a), (b) and (c), be rewritten, as regards truth value

$$F_{201}(F_{201}(F_{201}(F_{111}(P_{12}, P_{22}), F_{111}(P_{13}, P_{23})), F_{111}(P_{14}, P_{24})),$$
$$\times F_{202}(F_{202}(P_{22}, P_{23}), P_{24}))$$

A decision mechanism for this formula is thus obtained from that shown in Figure 7.2, by using suitable delay mechanisms.

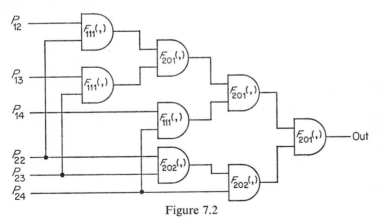

Figure 7.2

We note that

$$\lambda_{12} = 4 \quad \lambda_{13} = 4 \quad \lambda_{14} = 3 \quad \lambda_{22} = 4,3 \quad \lambda_{23} = 4,3 \quad \lambda_{24} = 3,2$$

Thus

$$\theta_{12} = 4 \quad \theta_{13} = 4 \quad \theta_{14} = 3 \quad \theta_{22} = 4 \quad \theta_{23} = 4 \quad \theta_{24} = 3$$

and

$$D = \max(2, 1, -1, 2, 1, -1) = 2$$

Hence we must delay the signals for P_{12}, P_{13}, P_{14} by $2-4+2, 2-4+3, 2-3+4$, i.e. by $0, 1, 3$ units of time respectively. The other three signals must each be delayed by two different lengths of time since $\lambda_{22}, \lambda_{23}, \lambda_{24}$ do not have unique values. The delays for signals corresponding to P_{22}, P_{23}, P_{24} are therefore 0 and 1, 1 and 2, 3 and 4 units respectively. The mechanism of Figure 7.2 is therefore replaced by that shown in Figure 7.3.

Examples 7A

1. Reduce the following formulae or, where appropriate, their negations, to disjunctive normal form and then rewrite, as far as truth values are concerned, the normal forms in terms of the functors $F_{201}(,)$, $F_{202}(,)$, $F_{111}(,)$:

    (a) $p \& q \lor (q \not\supset p \& r)$

    (b) $(p \equiv q \lor r) \supset p \& (r/q)$

    (c) $[\![p \supset (q \equiv r)]\!] \lor (q \equiv p \& r)$

    (d) $p \lor q \& (r \lor s)$

2. Repeat question 1 for functors

    $$F_{abc}(, \ldots,) \quad (a = 1, 2, \ldots; b = 0, 1, \ldots; c = 1, \ldots, a)$$

    obtaining solutions such that, in the corresponding decision mechanisms, each input is always connected to the final output through exactly two neurons in series.

3. Design, from neurons, an automaton with three inputs and one output such that the state of the output at time p corresponds to the proposition that between times $p-5$ and $p-2$ (both inclusive) all three inputs have been in states corresponding to the truth value T, the first such occurrence for the ith input being later than the first such occurrence for the $(i-1)$th input $(i = 2, 3)$.

4. Repeat question 3 using neurons corresponding to functors $F_{abc}(, \ldots,)$ for which $a + b \leqslant 2$, the output at time p of the previous question now being, for the least possible integer D, the output at time $p + D$.

5. Repeat questions 3 and 4 for the case where it is no longer required that the state of any input shall have corresponded to the truth value T in the relevant period; 'the first such occurrence' now being interpreted as 'the first such occurrence, if any'.

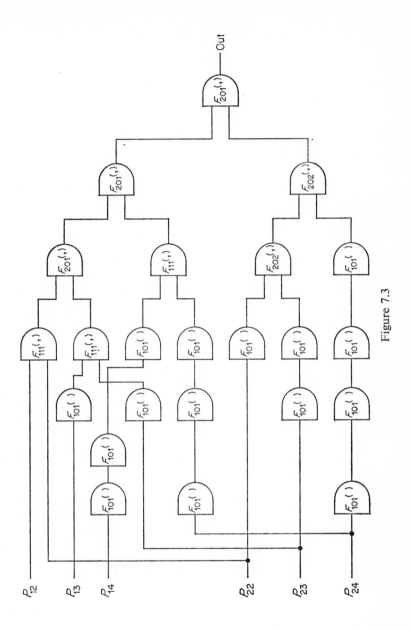

Figure 7.3

Solutions 7A

1. (a) If p, q, r take the truth value F, so does the given formula, which we shall therefore reduce (unnegated) to disjunctive normal form. We clearly obtain thereby the formula

$$p \And q \lor q \And \sim p \lor q \And \sim r$$

This takes the same truth value as the formula

$$F_{201}(F_{201}(F_{202}(p, q), F_{111}(q, p)), F_{111}(q, r))$$

2. (c) If p, q, r take the truth value F then the given formula takes the truth value T.

$$\sim \{[\![p \supset (q \equiv r)]\!] \lor (q \equiv p \And r)\}$$

$$=_{\mathrm{T}} [\![p \not\supset (q \equiv r)]\!] \And (q \not\equiv p \And r)$$

$$=_{\mathrm{T}} p \And (q \And \sim r \lor \sim q \And r) \And [\![q \And \sim (p \And r) \lor \sim q \And p \And r]\!]$$

$$=_{\mathrm{T}} (p \And q \And \sim r \lor p \And \sim q \And r) \And [\![q \And (\sim p \lor \sim r) \lor \sim q \And p \And r]\!]$$

$$=_{\mathrm{T}} (p \And q \And \sim r \lor p \And \sim q \And r) \And (q \And \sim p \lor q \And \sim r \lor \sim q \And p \And r)$$

$$=_{\mathrm{T}} p \And q \And \sim r \lor p \And \sim q \And r$$

(deleting absurdities after using a distributive law)

$$=_{\mathrm{T}} F_{201}(F_{212}(p, q, r), F_{212}(p, r, q))$$

(d) $p \lor q \And (r \lor s) =_{\mathrm{T}} p \lor q \And r \lor q \And s =_{\mathrm{T}} F_{301}(F_{101}(p), F_{202}(q, r), F_{202}(q, s))$

3. We note that the state of the third input cannot correspond to the truth value T at times $p - 5, p - 4$, since the state of the first input could not then correspond to this truth value sufficiently early. If the state of the third input so corresponds at time $p - 3$, the states of the second and first inputs must so correspond at times $p - 4$ (but not $p - 5$), $p - 5$ respectively. If the state of the third input so corresponds at time $p - 2$ (but not at time $p - 3$) then the states of the second and first inputs must so correspond either

(i) at times $p - 3$ (but not earlier), $p - 4$ respectively or

(ii) at times $p - 3$ (but not earlier), $p - 5$ respectively or

(iii) at times $p - 4$ (but not $p - 5$), $p - 5$ respectively.

The condition is therefore clearly given by the formula

$$\sim P_{35} \And \sim P_{34} \And (P_{33} \And P_{24} \And \sim P_{25} \And P_{15}$$

$$\lor P_{32} \And \sim P_{33} \And P_{23} \And \sim P_{24} \And \sim P_{25} \And P_{14}$$

$$\lor P_{32} \And \sim P_{33} \And P_{23} \And \sim P_{24} \And \sim P_{25} \And P_{15}$$

$$\lor P_{32} \And \sim P_{33} \And P_{24} \And \sim P_{25} \And P_{15})$$

which takes the truth value F when all its propositional variables take the truth value F. A reduction is given by

$$\sim P_{35} \,\&\, \sim P_{34} \,\&\, P_{33} \,\&\, P_{24} \,\&\, \sim P_{25} \,\&\, P_{15}$$
$$\vee \sim P_{35} \,\&\, \sim P_{34} \,\&\, P_{32} \,\&\, \sim P_{33} \,\&\, P_{23} \,\&\, \sim P_{24} \,\&\, \sim P_{25} \,\&\, P_{14}$$
$$\vee \sim P_{35} \,\&\, \sim P_{34} \,\&\, P_{32} \,\&\, \sim P_{33} \,\&\, P_{23} \,\&\, \sim P_{24} \,\&\, \sim P_{25} \,\&\, P_{15}$$
$$\vee \sim P_{35} \,\&\, \sim P_{34} \,\&\, P_{32} \,\&\, \sim P_{33} \,\&\, P_{24} \,\&\, \sim P_{25} \,\&\, P_{15}$$
$$=_{\mathrm{T}} F_{401}(F_{333}(P_{33}, P_{24}, P_{15}, P_{35}, P_{34}, P_{25}),$$
$$\times F_{353}(P_{32}, P_{23}, P_{14}, P_{35}, P_{34}, P_{33}, P_{24}, P_{25}),$$
$$\times F_{353}(P_{32}, P_{23}, P_{15}, P_{35}, P_{34}, P_{33}, P_{24}, P_{25}),$$
$$\times F_{343}(P_{32}, P_{24}, P_{15}, P_{35}, P_{34}, P_{33}, P_{25}))$$

7.3 Construction of finite automata from decision elements and delay units

In this section we shall revert to the consideration of ordinary decision elements and show how these may be used in combination with delay mechanisms to construct finite automata. We shall assume that the operating time of a decision element is negligible in comparison with the interval (of one unit of time) between clock pulses. One example of a construction of this type was given in Section 4.3 when we considered a serial adder.

The state of the final output (corresponding to the digit of the sum) at a given time depended on the states of the two inputs and also on the state of the other output (corresponding to the digit to be carried). Thus the truth value of P was determined not only by the truth values of p and q, but also by the truth value of Q_{i-1} and the state of the latter output (corresponding to Q_{i-1}) may be regarded as the internal state of the adder automaton.

More generally, if an automaton has n inputs whose states correspond to the truth values of formulae $P_1, \ldots, P_n$ and k internal states, we may regard these states as corresponding to the truth values of formulae $Q_1, \ldots, Q_m$ (built out of no sub-formulae other than formulae corresponding, at various times, past and present, to the inputs), where m is the least integer such that $2^m \geqslant k$ (since truth values may be assigned to $Q_1, \ldots, Q_m$ in 2^m ways, i.e. in at least k ways). To each of the k internal states we assign (in whatever way we find most convenient) a different ordered m-tuple of truth values of $Q_1, \ldots, Q_m$ respectively. The automaton contains decision mechanisms for the formulae $Q_1, \ldots, Q_m$ and the internal state of the automaton is then determined, as explained above, by the states of the m outputs of these mechanisms. The remaining $2^m - k$ assignments of truth values to $Q_1, \ldots, Q_m$ may (unless $k = 2^m$) be regarded as corresponding to $2^m - k$ further internal states of the automaton, these states not being assumed in practice. Thus we may lay down behaviour for the automaton when in these $2^m - k$ internal states in any way which we may find convenient.

Let us represent the states of the n inputs at time i by the truth values of the formulae $P_{1i}, ..., P_{ni}$ respectively and the states of the m outputs which correspond to the internal state of the automaton by the truth values of the formulae $Q_1, ..., Q_m$ respectively. We must therefore find formulae $R_i, Q_{1,i+1}, ..., Q_{m,i+1}$ built out of no sub-formulae other than $P_{1i}, ..., P_{ni}$; $Q_{1i}, ..., Q_{mi}$ such that the truth value of R_i represents the state of the output at time i and the truth values of $Q_{1,i+1}, ..., Q_{m,i+1}$ represent the internal state of the automaton at time $i+1$ ($i = 0, 1, ...$). Naturally $Q_{10}, ..., Q_{n0}$ all take the truth value F. (We note that the formulae P_i, Q_i of the serial adder become, in the present discussion, the formulae R_{i-1}, Q_{1i} respectively ($i = 1, 2, ...$), the value of m being 1.)

Having considered the example of the serial adder we shall now consider the application of the methods outlined above to the construction of an automaton with one input and one output such that if the state of the input represents, at all times, the binary digit 1 (and therefore the truth value T), the output's state represents, in succession, the digits 101010... and such that the interspersion of one or more digits of 0 into the input sequence causes a corresponding interspersion of zeros into the output sequence. (Thus, for example, the input sequence 10011101101111... would give rise to the output sequence 10001001001010....) The state of the input determines that of the output in two distinct ways.

(i) If α is even, and α inputs corresponding to 1 have in the past been supplied, the state of the output is the same as that of the input.

(ii) If, with the above notation, α is odd then the state of the output corresponds to 0.

Thus the automaton has two internal states to which we shall make the assignments of the truth values F, T respectively to a formula Q correspond. (We note that $k = 2$, so $m = 1$ and we omit the suffix 1 from Q. The formulae $Q_{10}, Q_{11}, ...$ are therefore renamed $Q_0, Q_1, ...$ respectively. We have by construction ensured that Q_0 takes the truth value F.) The internal state of the automaton at a given time differs from that one unit of time earlier if and only if the state of the input corresponded to the digit 1 (i.e. to the truth value T) at that earlier time. Thus we may take

$$Q_0 = f \quad R_i = P_i \not\supset Q_i \quad Q_{i+1} = Q_i \neq P_i \quad (i = 0, 1, ...)$$

We may (cf. Section 4.3) obtain inputs corresponding to the formulae Q_i by feeding inputs corresponding to Q_{i+1} into a delay of one unit ($i = 0, 1, ...$), since the new value of i (i.e. the value after one additional unit of time has elapsed) will be equal to the old value of $i+1$. Thus the automaton may be constructed as shown in Figure 7.4.

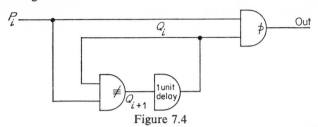

Figure 7.4

As a further example let us consider the design of an automaton which differs from the flip-flop of Section 4.8 (i.e. the R–S flip-flop) in that the state of the output does not correspond to the truth value T until the state of the set input has twice corresponded to this truth value. Simultaneous correspondence of the set and reset input states to the truth value T has the same effect as their simultaneous correspondence to the truth values F, T respectively. More precisely, the state of the output corresponds, at time p, to the truth value T if and only if there exist integers α, β, γ such that

$$0 \leqslant \beta < \gamma \leqslant \alpha \leqslant p,$$

the state of the reset input corresponds to the truth value F at times $p - \alpha, \ldots, p$ and the state of the set input corresponds to the truth value T at times $p - \alpha + \beta, \, p - \alpha + \gamma$.

Let the states of the set and reset inputs and of the output at time i correspond to the truth values of the formulae P_{1i}, P_{2i}, R_i respectively ($i = 0, 1, \ldots$). Thus the truth value of R_p is determined by those of P_{1p} and P_{2p} in one of three possible ways ($p = 0, 1, \ldots$).

(i) If there exist integers α, β, γ such that $0 \leqslant \beta < \gamma \leqslant \alpha \leqslant p-1$ and $P_{1,p-1-\alpha+\beta}, P_{1,p-1-\alpha+\gamma}, P_{2,p-1-\alpha}, \ldots, P_{2,p-1}$ take the truth values $T, T, F, \ldots, F$ respectively then

$$R_p =_T \sim P_{2p}$$

(ii) If there exist integers α, β such that $0 \leqslant \beta \leqslant \alpha \leqslant p-1$, the formulae $P_{1,p-1-\alpha}, \ldots, P_{1,p-1-\alpha+\beta-1}, P_{1,p-1-\alpha+\beta+1}, \ldots, P_{1,p-1}, P_{2,p-1-\alpha}, \ldots, P_{2,p-1}$ take the truth value F and $P_{1,p-1-\alpha+\beta}, P_{2,p-2-\alpha}$ take the truth value T ($P_{2,-1}$ being regarded as taking the truth value T) then

$$R_p =_T P_{1p} \not\Rightarrow P_{2p}$$

(iii) If there exists an integer α such that

$$0 \leqslant \alpha \leqslant p \quad \text{and} \quad P_{1,p-\alpha}, \ldots, P_{1,p-1}, \; P_{2,p-1-\alpha}$$

take the truth values $F, \ldots, F, T$ respectively then R_p takes the truth value F ($P_{2,-1}$ again being regarded as taking the truth value T). The automaton must therefore be capable of assuming three distinct internal states corresponding to (i), (ii) and (iii). Thus, with the notation of the second paragraph of this section, $k = 3$ and it follows at once that $m = 2$.

The state corresponding to (iii), being the initial state, must of course be represented by the assignment of the truth value F to Q_{1i}, Q_{2i}. Let us assign the truth values T, F to Q_{1i}, Q_{2i} respectively in the redundant case, and the truth values T/F, T to Q_{1i}, Q_{2i} respectively in the case corresponding to (i)/(ii). Hence we obtain the following truth table for R_p, for convenience assigning the truth value F to R_p in the redundant case, which we shall name (iv).

	P_{1p}	P_{2p}	Q_{1p}	Q_{2p}	R_p
(i) $R_p =_T \sim P_{2p}$	T	T	T	T	F
	F	T	T	T	F
	T	F	T	T	T
	F	F	T	T	T
(ii) $R_p =_T P_{1p} \not\supset P_{2p}$	T	T	F	T	F
	F	T	F	T	F
	T	F	F	T	T
	F	F	F	T	F
(iii), (iv)	T or F	T or F	T or F	F	F

R_p takes the truth value F in all 8 cases.

We may therefore take

$$R_p = (P_{1p} \vee Q_{1p}) \,\&\, Q_{2p} \not\supset P_{2p}$$

Proceeding similarly, we obtain the following truth tables for $Q_{1,p+1}, Q_{2,p+1}$, treating (iv), for convenience, in the same way as (iii).

	P_{1p}	P_{2p}	Q_{1p}	Q_{2p}	$Q_{1,p+1}$	$Q_{2,p+1}$
(iii), (iv)	F	F	T or F	F	F	F
	F	T	T or F	F	F	F
	T	F	T or F	F	F	T
	T	T	T or F	F	F	F
(ii)	T	T	F	T	F	F
	F	T	F	T	F	F
	T	F	F	T	T	T
	F	F	F	T	F	T
(i)	T	T	T	T	F	F
	F	T	T	T	F	F
	T	F	T	T	T	T
	F	F	T	T	T	T

Thus we may take

$$Q_{1,p+1} = R_p \quad Q_{2,p+1} = [\sim P_{2p}, Q_{2p}, P_{1p} \dotplus P_{2p}]$$

and the circuit is as shown in Figure 7.5.

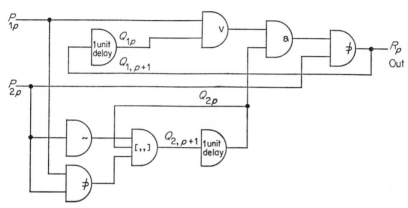

Figure 7.5

As an example of an automaton with two outputs let us consider a device with three inputs whose states at time i correspond to the values of the digits x_i, u_i, v_i and such that the states of the outputs at this time correspond to the values of the digits y_i, z_i ($i = 0, 1, \ldots$). Suppose that

$$y_i = 1 \quad \text{if and only if} \quad x_i = x_{i-1} = 1 \quad \text{and} \quad u_i \neq v_i \quad (i = 0, 1, \ldots)$$

and that

$$z_i = 1 \quad \text{if and only if} \quad x_i = x_{i-1} = 0 \quad \text{and} \quad u_i = v_i \quad (i = 0, 1, \ldots),$$

x_{-1} and y_{-1} being regarded as 0. Thus we require two internal states corresponding to the values of x_{i-1}. We therefore assign to Q_i the truth value T if and only if $x_{i-1} = 1$. (Since $x_{-1} = 0$, this rule gives Q_0 the truth value F as required in all cases.) Hence, if the outputs relating to the values of y_i, z_i correspond to the formulae R_i, S_i we may take

$$R_i = (P_{1i} \& Q_i) \dotplus (P_{2i} \equiv P_{3i})$$

$$S_i = (P_{1i} \downarrow Q_i) \& (P_{2i} \equiv P_{3i})$$

$$Q_{i+1} = P_{1i}$$

and use the circuit shown in Figure 7.6. We note that the formula $P_{2i} \equiv P_{3i}$ is a sub-formula both of R_i and of S_i and we can therefore effect an economy in the use of hardware. Such economies are discussed in detail by Kobrinskii and Trakhtenbrot.[30]

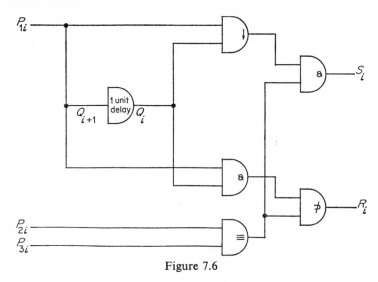

Figure 7.6

As a further example of an automaton with three inputs and two outputs let us, with the above notation, consider the case where, for $i = 0, 1, ...,$

$$y_i = 1 \quad \text{if and only if} \quad x_i = 1, \quad x_{i-1} = 0 \quad \text{and} \quad u_i \ne v_{i-1}$$

and

$$z_i = 1 \quad \text{if and only if} \quad x_i = 0, \quad x_{i-1} = 1 \quad \text{and} \quad u_i = v_{i-1}$$

We require four internal states corresponding to the four possible pairs of values of x_{i-1}, v_{i-1}. Let us assign to Q_{1i} (Q_{2i}) the truth value T if and only if x_{i-1} $(v_{i-1}) = 1$. We may therefore take

$$Q_{1,i+1} = P_{1i} \quad Q_{2,i+1} = P_{3i} \quad R_i = (P_{1i} \not\equiv Q_{1i}) \not\equiv (P_{2i} \equiv Q_{2i})$$

$$S_i = (Q_{1i} \not\equiv P_{1i}) \,\&\, (P_{2i} \equiv Q_{2i})$$

Examples 7B

1. Design, drawing a block diagram and using decision elements and delay units, an automaton with two inputs and one output such that, with the notation of the fourth paragraph of Section 7.3, if x_i $(y_i) = 1$ or 0 according as P_{1i} (P_{2i}) takes the truth value T or the truth value F $(i = 0, 1, ...)$, then R_p takes the truth value T if and only if

$$\sum_{i=0}^{p} 2^{p-i} x_i > \sum_{i=0}^{p} 2^{p-i} y_i \quad (p = 0, 1, ...)$$

2. Repeat question 1 replacing the inequality by

(i)
$$\sum_{i=0}^{p} 2^{p-i} x_i > \sum_{i=0}^{p} 2^{p-i} y_i + 2$$

(ii)
$$\sum_{i=0}^{p} 2^{p-i} x_i > \sum_{i=0}^{p} 2^{p-i} y_i + 5$$

(iii)
$$\sum_{i=0}^{p} 2^{p-i} x_i > 3 \sum_{i=0}^{p} 2^{p-i} y_i$$

3. Design, drawing a block diagram and using decision elements and delay units, an automaton with one input and one output such that
   (i) if the inputs correspond, in succession, to the digits 101010 ... then the outputs correspond, in succession, to the digits 110110110 ...,
   (ii) if the inputs correspond, in order, at times $0, ..., N-1$ to the first N digits of the sequence 101010 ... but the input at time N does not so correspond, then the outputs correspond, in order, at times $0, ..., N-1$ to the first N digits of the sequence 110110110 ... and the outputs at times $N, N+1, ...$ all correspond to the digit 0 ($N = 0, 1, ...$).
4. Repeat question 3, interchanging the roles of the two sequences.
5. Repeat question 1 for a two-output automaton corresponding to the following pairs of conditions, where

$$X_p = \sum_{i=0}^{p} 2^{p-i} x_i \quad Y_p = \sum_{i=0}^{p} 2^{p-i} y_i$$

(i) $\qquad X_p > Y_p + 1 \quad X_p < Y_p$

(ii) $\qquad X_p > 2Y_p \quad X_p < 3Y_p$

Solutions 7B

1. Let

$$\sum_{i=0}^{p} 2^{p-i} x_i = X_p \quad \sum_{i=0}^{p} 2^{p-i} y_i = Y_p \quad (p = 0, 1, ...)$$

Then $X_p > Y_p$ if and only if either

(i) $\qquad\qquad X_{p-1} > Y_{p-1}$

or

(ii) $\qquad x_p = 1, \quad y_p = 0 \quad$ and not $\quad Y_{p-1} > X_{p-1}$

($p = 0, 1, ...$; X_{-1} and Y_{-1} being regarded as 0). Thus we require three internal states corresponding to the relations

$$X_{p-1} > Y_{p-1}, \quad X_{p-1} = Y_{p-1}, \quad X_{p-1} < Y_{p-1}$$

and we shall represent these by the formulae Q_{1i}, Q_{2i} ($i = 0, 1, ...$). Q_{1i} takes the truth value T if and only if $X_{i-1} > Y_{i-1}$ and Q_{2i} takes the truth value T if and only if $X_{i-1} < Y_{i-1}$, so that the redundant state corresponds to the case where Q_{1i}, Q_{2i} both take the truth value T ($i = 0, 1, ...$). Hence we may take, for $p = 0, 1, ...$,

$$R_p = Q_{1p} \vee (P_{1p} \not\supset P_{2p} \vee Q_{2p})$$

$$Q_{1, p+1} = R_p$$

$$Q_{2, p+1} = Q_{2p} \vee (P_{2p} \not\supset P_{1p} \vee Q_{1p})$$

2. (i) We define $X_{-1}, X_0, X_1, \ldots, Y_{-1}, Y_0, Y_1, \ldots$ as in the previous solution. Hence

   (i) if $X_{p-1} > Y_{p-1} + 2$ then $X_p > Y_p + 2$

  (ii) if $X_{p-1} = Y_{p-1} + 2$ then $X_p > Y_p + 2$

 (iii) if $X_{p-1} = Y_{p-1} + 1$ then

      (a) if $x_p = 1, y_p = 0$ then $x_p > Y_p + 2$

      (b) if $x_p = y_p$ then $X_p = Y_p + 2$

      (c) if $x_p = 0, y_p = 1$ then $X_p = Y_p + 1$

 (iv) if $X_{p-1} = Y_{p-1}$ then

      (a) if $x_p = 1, y_p = 0$ then $X_p = Y_p + 1$

      (b) if $x_p = y_p$ then $X_p = Y_p$

      (c) if $x_p = 0, y_p = 1$ then $X_p < Y_p$

  (v) if $X_{p-1} < Y_{p-1}$ then $X_p < Y_p$

We therefore require five internal states corresponding to the above five relations connecting X_{p-1} and Y_{p-1}, so that $k = 5$ and $m = 3$. Since $X_{-1} = Y_{-1} (= 0)$ the initial internal state corresponds to (iv) and this must, of course, be represented by assigning the truth value F to Q_1, Q_2, Q_3. We shall represent the five states by assigning truth values to Q_1, Q_2, Q_3 as shown below, the fourth of the five assignments being, of course, as stated above.

	Q_1	Q_2	Q_3
(i)	T	T	F
(ii)	F	T	F
(iii)	T	F	F
(iv)	F	F	F
(v)	T	T	T
Redundant states	F	T	T
	T	F	T
	F	F	T

Conditions (i)–(v) thus provide the following partial truth tables.

	P_{1p}	P_{2p}	Q_{1p}	Q_{2p}	Q_{3p}	R_p	$Q_{1,p+1}$	$Q_{2,p+1}$	$Q_{3,p+1}$
(i)	T or F	T or F	T	T	F	T	T	T	F
(ii)	T or F	T or F	F	T	F	T	T	T	F
(iii)	T	F	T	F	F	T	T	T	F
	F	F	T	F	F	F	F	T	F
	T	T	T	F	F	F	F	T	F
	F	T	T	F	F	F	T	F	F
(iv)	T	F	F	F	F	F	T	F	F
	F	F	F	F	F	F	F	F	F
	T	T	F	F	F	F	F	F	F
	F	T	F	F	F	F	T	T	T
(v)	T or F	T or F	T	T	T	F	T	T	T

Hence we may take

$$R_p = Q_{2p} \vee (P_{1p} \,\&\, Q_{1p} \,\dot{\supset}\, P_{2p}) \,\dot{\supset}\, Q_{3p}$$
$$Q_{1,p+1} = Q_{2p} \vee (P_{1p} \not\equiv P_{2p})$$
$$Q_{2,p+1} = Q_{2p} \vee \llbracket Q_{1p} \equiv (P_{2p} \,\dot{\supset}\, P_{1p}) \rrbracket$$
$$Q_{3,p+1} = Q_{3p} \vee \llbracket P_{2p} \,\dot{\supset}\, D_3(P_{1p}, Q_{1p}, Q_{2p}) \rrbracket$$

3. Since $n = 1$ we may represent the state of the input at time i by the truth value of the formula P_i ($i = 0, 1, ...$). Thus, for $p = 0, 1, ...$ (regarding $P_0, ..., P_{-1}$ as corresponding to the first zero members of any sequence)

  (i) if $p \equiv 0 \pmod 6$ and the truth values of $P_0, ..., P_{p-1}$ correspond to the first p members of the sequence 101010 ... then, if P_p takes the truth value T, R_p takes the truth value T, $p + 1 \equiv 1 \pmod 6$ and the truth values of $P_0, ..., P_p$ correspond to the first $p + 1$ members of the above sequence. If P_p takes the truth value F then R_p takes the truth value F and the truth values of $P_0, ..., P_p$ do not so correspond.

  (ii)–(vi) are obtained from (i) by replacing the congruences (mod 6) and the first truth values of P_p, R_p respectively as shown below, the second truth value of P_p always differing from the first.

  (ii) $p \equiv 1, \quad p + 1 \equiv 2, F, T$

  (iii) $p \equiv 2, \quad p + 1 \equiv 3, T, F$

  (iv) $p \equiv 3, \quad p + 1 \equiv 4, F, T$

  (v) $p \equiv 4, \quad p + 1 \equiv 5, T, T$

  (vi) $p \equiv 5, \quad p + 1 \equiv 0, F, F$

  (vii) If the truth values of $P_0, ..., P_{p-1}$ do not correspond to the first p members of the sequence 101010 ... then R takes the truth value F and the truth values of $P_0, ..., P_p$ do not correspond to the first $p + 1$ members of the sequence 101010 ... .

  Thus we require seven internal states, relating to the conditions (i)–(vii), and the state relating to (i) corresponds to the assignment of the truth value F to Q_1, Q_2, Q_3. (Since $k = 7$ it follows at once that $m = 3$.) Thus, for suitable representations of the other seven internal states (including the redundant state), we obtain the following partial truth tables.

	P_p	Q_{1p}	Q_{2p}	Q_{3p}	R_p	$Q_{1,p+1}$	$Q_{2,p+1}$	$Q_{3,p+1}$
(i)	T	F	F	F	T	T	F	F
	F	F	F	F	F	T	T	T
(ii)	F	T	F	F	T	F	T	F
	T	T	F	F	F	T	T	T
(iii)	T	F	T	F	F	T	T	F
	F	F	T	F	F	T	T	T
(iv)	F	T	T	F	T	F	F	T
	T	T	T	F	F	T	T	T
(v)	T	F	F	T	T	T	F	T
	F	F	F	T	F	T	T	T
(vi)	F	T	F	T	F	F	F	F
	T	T	F	T	F	T	T	T
(vii)	T or F	T	T	T	F	T	T	T

Hence we may take

$$R_p = [\![(P_p \not\equiv Q_{1p}) \not\supset Q_{2p} \vee Q_{3p}]\!] \vee C_3(P_p \not\equiv Q_{1p}, P_p \not\equiv Q_{2p}, P_p \equiv Q_{3p})$$

and the formulae $Q_{i,p+1}$ $(i = 1, 2, 3)$ may be found similarly. Alternatively we may (using a distributive law and the above result) take

$$R_p = (P_p \not\equiv Q_{1p}) \,\&\, [\![Q_{2p} \vee Q_{3p} \supset (P_p \not\equiv Q_{2p}) \,\&\, (P_p \, Q_{3p})]\!]$$

5. (i) Let R_p take the truth value T if and only if $X_p > Y_p + 1$ and S_p take the truth value T if and only if $X_p < Y_p$. We note that the following implications hold.

(i) If $X_{p-1} > Y_{p-1} + 1$ then $X_p > Y_p + 1$
(ii) If $X_{p-1} = Y_{p-1} + 1$ then
    (a) if $x_p \geqslant y_p$ then $X_p > Y_p + 1$
    (b) if $x_p = 0, y_p = 1$ then $X_p = Y_p + 1$
(iii) If $X_{p-1} = Y_{p-1}$ then
    (a) if $x_p = 1, y_p = 0$ then $X_p = Y_p + 1$
    (b) if $x_p = y_p$ then $X_p = Y_p$
    (c) if $x_p = 0, y_p = 1$ then $X_p < Y_p$
(iv) If $X_{p-1} < Y_{p-1}$ then $X_p < Y_p$

Thus we require four internal states and, for suitable representations of these, we obtain the partial truth tables shown below. Hence we may take

$$R_p = Q_{1p} \,\&\, [\![Q_{2p} \vee (P_{2p} \supset P_{1p})]\!]$$

$$S_p = Q_{2p} \vee (P_{2p} \not\supset P_{1p}) \not\supset Q_{1p}$$

$$Q_{1,p+1} = Q_{1p} \vee (P_{1p} \not\supset P_{2p} \vee Q_{2p})$$

$$Q_{2,p+1} = R_p \vee S_p$$

	P_{1p}	P_{2p}	Q_{1p}	Q_{2p}	R_p	S_p	$Q_{1,p+1}$	$Q_{2,p+1}$
(i)	T or F	T or F	T	T	T	F	T	T
(ii)	T	T	T	F	T	F	T	T
	T	F	T	F	T	F	T	T
	F	F	T	F	T	F	T	T
	F	T	T	F	F	F	T	F
(iii)	T	F	F	F	F	F	T	F
	T	T	F	F	F	F	F	F
	F	F	F	F	F	F	F	F
	F	T	F	F	F	T	F	T
(iv)	T or F	T or F	F	T	F	T	F	T

References

1. C. E. Shannon, 'A symbolic analysis of relay and switching circuits', *Trans. A.I.E.E.*, **57**, 713 (1938).
2. D. B. McCallum and J. B. Smith, 'Mechanized reasoning—Logical computers and their design', *Electronic Engineering*, **23**, 126 (1951).
3. J. Łukasiewicz, *Elements of Mathematical Logic*, Pergamon, Oxford, 1963.
4. D. Hilbert and W. Ackermann, *Principles of Mathematical Logic*, Chelsea, New York, 1950.
5. J. E. Parton and A. Rose, 'The universal decision element', *Trans. Soc. Instrum. Tech.*, **13**, 177 (1961).
6. A. Church, 'Conditioned disjunction as a primitive connective for the propositional calculus', *Portugal. Math.*, **7**, 87 (1948).
7. E. L. Post, *The 2-valued Iterative Systems of Mathematical Logic*, Annals of Mathematics Studies, No. 5, Princeton, 1941.
8. A. J. Khambata, *Introduction to Integrated Semiconductor Circuits*, Wiley, New York, 1963.
9. R. S. Ledley, *Digital Computer and Control Engineering*, McGraw-Hill, New York, 1960.
10. W. E. Wickes, *Logic Design with Integrated Circuits*, Wiley, New York, 1968.
11. R. L. Goodstein, *Boolean Algebra*, Pergamon, Oxford, 1963.
12. C. E. Shannon and J. McCarthy (eds.), *Automata Studies*, Annals of Mathematics Studies, No. 34, Princeton, 1956.
13. D. B. McCallum and J. B. Smith, 'Feedback logical computers', *Electron. Engng*, **23**, 458 (1951).
14. A. Rose, 'Sur les éléments universels de décision', *Comptes rendus (Paris)*, **244**, 2343 (1957).
15. B. Sobociński, 'On a universal decision element', *J. Comptg Systs*, **1**, 71 (1953).
16. S. Leśniewski, 'Grundzüge eines neuen Systems der Grundlagen der Mathematik', *Fund. Math.*, **14**, 1 (1929).
17. E. C. Mihailescu, 'Recherches sur la négation et l'équivalence dans le calcul des propositions', *Ann. Sci. Univ. Jassy*, **23**, 369 (1937).
18. A. Rose, 'A high speed parallel adder', *Z. für Math. Log.*, **5**, 240 (1959).
19. T. Kilburn, D. B. G. Edwards and D. Aspinall, 'Parallel addition in digital computers: A new fast "carry" circuits', *The Institution of Electrical Engineers, Measurement and Control Section, Specialist Discussion Meetings on New Digital Computer Techniques*, 16–17 February 1959, p. 17.
20. A. Rose, 'The use of universal decision elements as flip-flops', *Z. für Math. Log.*, **4**, 169 (1958).
21. A. Rose, 'Applications of logical computers to the construction of electrical control tables for signalling frames', *Z. für Math. Log.*, **4**, 222 (1958).

22. W. V. Quine, 'The problem of simplifying truth-functions', *Amer. Math. Monthly*, **59**, 521 (1952).
23. M. Phister, *Logical Design of Digital Computers*, Wiley, New York, 1958.
24. J. P. Roth, 'Algebraic topological methods for the synthesis of switching systems, 1', *Trans. Am. Math. Soc.*, **88**, 301 (1958).
25. G. Birkhoff and S. MacLane, *A Survey of Modern Algebra*, 3rd ed., Macmillan, New York, 1965.
26. D. A. Huffman, 'The synthesis of sequential switching networks', *J. Franklin Inst.*, **257**, 161, 275 (1954).
27. G. H. Mealey, 'A method for synthesizing sequential circuits', *Bell System Tech. J.*, **34**, 1045 (1955).
28. S. H. Caldwell, *Switching Circuits and Logical Design*, Wiley, New York, 1958.
29. A. M. Turing, 'On computable numbers', *Proc. Lond. Math. Soc.*, II, **42**, 230 (1936).
30. N. E. Kobrinskii and B. A. Trakhtenbrot, *Introduction to the Theory of Finite Automata*, North Holland, Amsterdam, 1965.

Author Index

Reference numbers are in square brackets and the page on which the author's work is quoted in full is given in italics.

175

Subject Index

176